An Introduction to Spatial Data Analysis

An Introduction to Spatial Data Analysis

Remote Sensing and GIS with Open Source Software

Martin Wegmann, Jakob Schwalb-Willmann
and Stefan Dech

DATA IN THE WILD SERIES

Pelagic Publishing | www.pelagicpublishing.com

Published by Pelagic Publishing
PO Box 874
Exeter
EX3 9BR
UK

www.pelagicpublishing.com

An Introduction to Spatial Data Analysis:
Remote Sensing and GIS with Open Source Software

ISBN 978-1-78427-212-8 (Hbk)
ISBN 978-1-78427-213-5 (Pbk)
ISBN 978-1-78427-214-2 (ePub)
ISBN 978-1-78427-215-9 (ePDF)

A CIP record for this book is available from the British Library

Cover image: Modified data of elevation values within the study area

Printed and bound in India by Replika Press Pvt. Ltd.

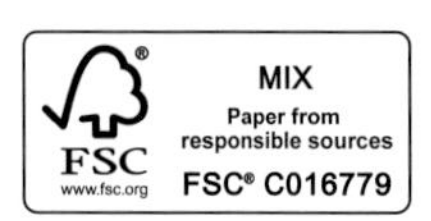

Contents

Preface

Welcome to *Introduction to Spatial Data Analysis*! In this book, you will learn the essentials of spatial data handling using the free and open-source geographic information system (GIS) software QGIS. In addition, you are invited to take your first steps in using the R programming language. This book covers the fundamentals of spatial data handling and analysis. It will empower you to turn data acquired in the field (in situ/ground truth data) into actual spatial data, process and analyse different types of spatial data and interpret the data and results. After finishing this book, you will be able to address questions such as 'What is the distance to the border of the protected area?', 'Which points are located close to a road?' and 'Which fraction of land cover types exist in my study area?' using different software and techniques.

Even though this book is clearly aimed at the beginner level of working with spatial data, it assumes a sound background in environmental sciences and scientific methods and a good idea why and in what contexts spatial analyses may be useful and can be applied. The approaches introduced in the book are aimed at general environmental science requirements. As the book's study region to demonstrate our analyses, we chose an area in the Steigerwald nature park in the German uplands. There, we have an existing collaboration with the local ecological research station.

This book will guide you through the first steps you need to consider before collecting and assembling any spatial data, followed by commonly needed spatial data manipulation and analysis methods applied in the environmental sciences. You will learn how to get field data into a spatial format using a GIS, how to modify it and create spatial queries to derive new information. This will be done using QGIS, a software program with a graphical user interface (GUI). This way, you can start without using any code. However, the book ends with an introduction to and first steps using the R programming language. Thus, you will learn to do the analyses conducted in QGIS using R commands. This will allow you to get started writing your own scripts for spatial data analysis.

In addition, we explain how to create suitable visualizations of spatial data and discuss examples of good and bad maps and graphs. Thus, we cover all topics needed to get started working in the spatial domain and achieving your first meaningful results.

The first chapters will guide you through the basic spatial data analysis theory and practice along with giving you a first idea of working with software designed to handle spatial data. Various examples are explained and executed. These sample operations aim to outline possibilities and should not be considered as strict guidelines for your specific research project. For this, we will use artificial data and examples to teach you how to get started with spatial data collection, management and analysis. In subsequent chapters, actual data for a specific study area are used and discussed.

The book has four parts:

- Part I: data acquisition and preparation and map creation.
- Part II: spatial field data acquisition and creation of auxiliary data.

- Part III: data analysis and creation of new spatial information.
- Part IV: introduction to the topics covered before using R code.

By the time you finish working through this book, you will have learnt:

- What to consider before you collect data in situ.
- How to work with spatial data collected in situ.
- The difference between raster and vector data.
- How to acquire further vector and raster data.
- How to create relevant environmental information.
- How to combine and analyse in situ and remote sensing data.
- How to create useful maps for field work and presentations.
- How to use QGIS and R for spatial analysis.
- How to develop your own analysis script.

0.1 Who this book is for

As the title of the book implies, its contents address novice spatial data users and do not assume any prior knowledge about spatial data itself or practical experience working with such data sets. Unlike *Remote Sensing and GIS for Ecologists: Using Open Source Software* (http://book.ecosens.org), which assumes intermediate level background knowledge and focuses mainly on the use of R for all necessary tasks, this book starts with the basics of working with spatial data and provides guidelines on how to start without using any code.

The book is meant to impart the essential knowledge needed to work with spatial data. This captures a wide range of raster and vector data collection, conversion, generation and analysis. It starts right at the beginning with the data collection of either existing spatial data or using a global positioning system (GPS), transferring GPS data into a GIS and conducting all the necessary further steps. As a result, this book can benefit ecologists, geographers and any environmental scientists or practitioners who need to collect, visualize and analyse spatial data.

Beginners in spatial data handling will gain a basic understanding of the functionality of QGIS and R. Intermediate or advanced R (coding) users will likely be happier with *Remote Sensing and GIS for Ecologists*, which starts with R and provides all the necessary R code to analyse spatial data for various purposes. That book also covers more advanced spatial data analysis approaches, such as texture analysis or species distribution modelling.

0.2 What you need to use this book

To learn spatial data handling in QGIS and R, we highly recommend that you install the software yourself and redo the exercises on your computer system. In the following chapters, the QGIS software is explained and introduced in detail. R and the integrated development environment RStudio are introduced later in the book. The software can be installed from https://qgis.org/en/site/, https://www.r-project.org and https://rstudio.com, respectively. For this book, we have used the latest versions, that is, QGIS v3.8, R v3.6.x and RStudio v1.1.x. Later versions should provide the same interfaces and functionalities. We have used Linux to run the software and take screenshots for the book. Thus, the software interfaces may look slightly different in Windows or macOS distributions.

0.2.1 Data

Most data sets used in the book can be acquired and created by you by following the instructions provided. A few data sets are provided by us as a basis for the analysis. This includes example data used in Chapter 1 and the 'area of interest', defining the study area of the analysis, starting with Chapter 2.

Nevertheless, we have decided to provide the data acquired and created within each chapter to allow you to cross-check your own results. Some chapters may use the results of previous chapters. Additionally, we provide data sets that are not used in subsequent chapters. Figure 0.1 gives an overview of all the data sets provided with the book, sorted according to chapter. However, we highly recommend that you build up your own data structure step by step by following the tasks provided and only use the data sets provided to compare your results or when you are stuck with a task. Note that in the chapters that introduce R, you need to adapt the file paths according to your own data structure.

Please familiarize yourself with the data structure of the downloaded data or decide on your own data storage structure.

0.2.2 Software programs

QGIS is a platform-independent, free, open-source GIS. QGIS is a program like proprietary GIS packages and offers a GUI with a variety of common GIS functionalities. It can be used quite intuitively, and many common and more complex tasks are easily feasible using QGIS. If you are a beginner, any GIS and coding task might be challenging at first. All scientific software packages are usually complex and cannot be mastered within a few days; that is also true for QGIS and R. For the analysis on the command line we use R, a statistical programming environment with a great variety of functions. If you have never worked using the command line before, you may be intimidated by the thought of writing program code instead of clicking on a button with your mouse in a GUI. Hopefully, you will feel encouraged to write your analysis in a coding environment after reading this book. Normally, it takes our students up to one year of coursework to be entirely comfortable with coding environments like R, so please do not expect miracles. You will have to work continuously on your coding skills and fight any frustrating bits and pieces you may encounter along the way. But you will also be rewarded once you have managed to solve problems and finish off your script.

Coding might be an option for you if you have encountered one of the following situations:

- First, when you are done with the analysis, you have a complete protocol of what you did. Have you ever tried to remember the steps you took half a year ago within a GUI when you did your analysis?
- Second, because of the script you wrote, you can rerun the full analysis again with a single click without having to click through hundreds of GUI options. Have you ever discovered an error in your input data forcing you to repeat the whole analysis?
- Third, it is easy to share your script with other researchers. This is an important step towards reproducible research. Have you ever tried to explain that complete GUI workflow to your colleagues, for example, 'First you click there, then you click here, then …'?

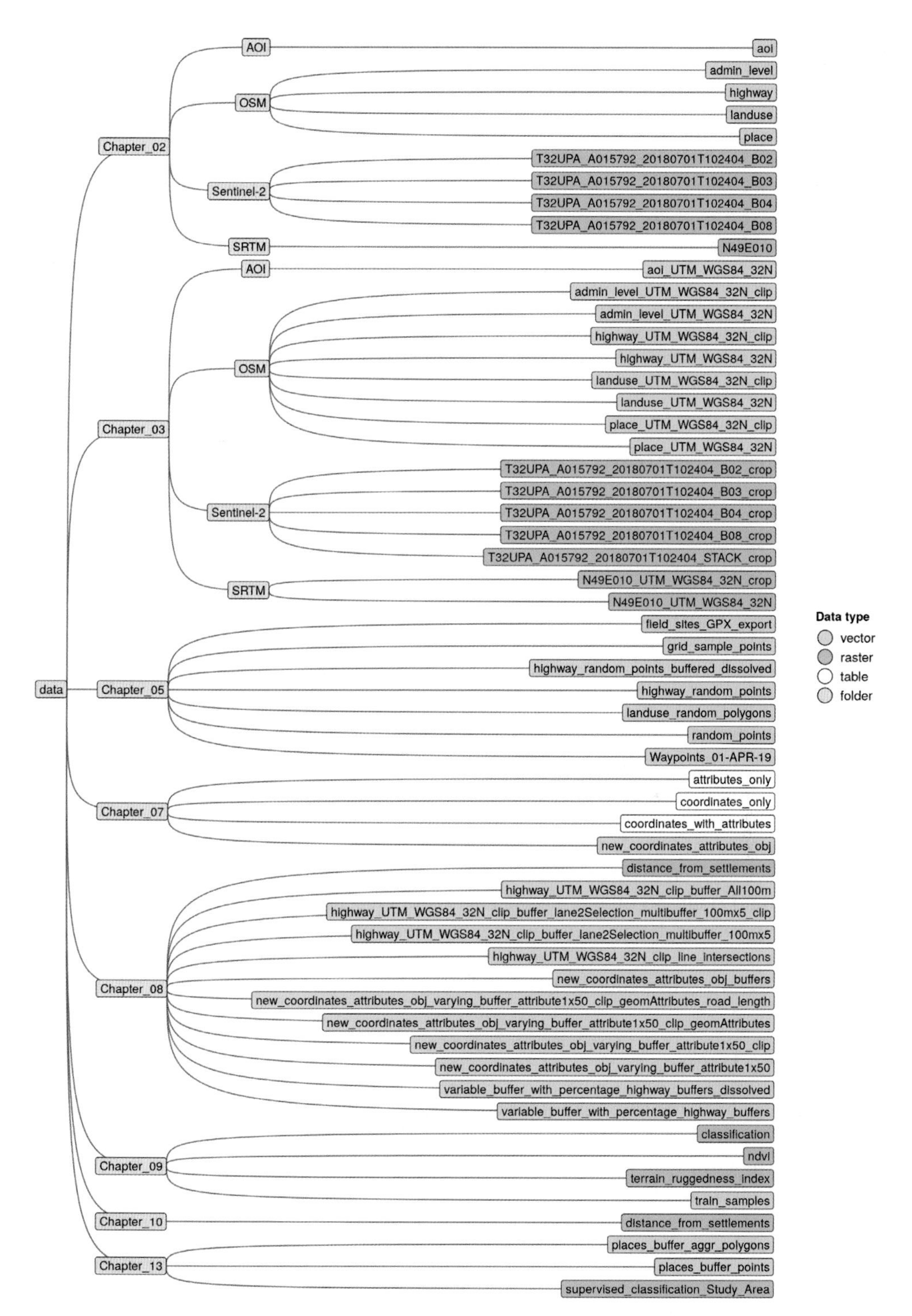

Figure 0.1 An overview of the data sets created and used in this book, split according to chapters and coloured by type.

R is a high-level interpreter language. This means that R code does not need to be compiled after finishing working on it but can be executed line by line while working on it. This makes R very user-friendly since you can formulate, write, run and check an analysis step directly before continuing with the next. This also means that you do not have to work with the fine-grained details of compiled language programming like C++ or Fortran. Rather, R provides an abstracted set of functions that are often easy to read and understand, even for non-programmers. The R coding environment at the end of this book aims to be a smooth transition to the more R-focused book *Remote Sensing and GIS for Ecologists*.

Because it might confuse some readers, let us stress this once more: our user interfaces might differ slightly if you do not work in Linux. In addition, some icons or the look and feel of QGIS might change with newer versions of the software; however, the general functions will remain the same.

0.3 Bugs, problems and challenges

We have tried to provide you with working examples. However, anticipating all potential errors, the different behaviours of operating systems, changes in software functionality or command syntax is rather difficult. Therefore, you will have to learn how to interpret error messages. In the beginning, this can be challenging. If you do not understand the error message, it can be extremely helpful to search for solutions online. Chances are, you are not the first one to encounter it! This will help you develop an understanding of what they mean. Please read the error messages carefully; usually, errors can be solved within minutes that way.

Both QGIS and R have very active email lists where all kinds of questions can be asked, and answers obtained. If you write to an email list, make sure to provide a small replicable example so that colleagues on the list can redo the example and experience the same challenges. Please remember that all list subscribers are volunteers. Their main job is not to help you, but they like to do so on top of their normal workload as scientists. Therefore, we ask you to do your best to solve your problem on your own and remember to be kind and polite to others. Please check the Web pages of the respective software programs for further details on email lists or other communication channels for help.

For discussions and questions on approaches or technical details, we have established an email list (https://lists.uni-wuerzburg.de/mailman/listinfo/rs-ecology).

0.4 Info boxes

Info boxes can be found throughout the chapters. These are either **R command info boxes**, where the corresponding R commands for the QGIS operations are listed, or **QGIS info boxes**, where additional tips and tricks using QGIS are provided. Lastly, in **task boxes**, we suggest additional tasks to try out in either QGIS or R.

These info boxes do not provide crucial information to understand spatial data handling; rather, they aim to link the different parts and encourage you to explore new functions.

0.5 Further reading

More details on QGIS, QGIS map design or R are provided online and in various books or manuals:

- QGIS provides good documentation on its various functionalities, which are available at https://qgis.org/en/docs/index.html.
- A list of QGIS books are available at https://qgis.org/en/site/forusers/books/.
- Anita Graser and Gretchen Peterson's *QGIS Map Design*, Locate Press, contains great examples of maps created with QGIS.
- Some QGIS teaching material can be found at https://qgis.org/en/site/forusers/trainingmaterial/index.
- Andy Field, Jeremy Miles and Zoë Field's *Discovering Statistics Using R*, SAGE Publications Ltd, and the examples within, are liked by many.
- Mark Dale and Marie-Josée Fortin's *Spatial Analysis: a Guide for Ecologists*, Cambridge University Press, provides a good overview of spatial statistics.
- Michael Crawley's *The R Book* is a standard book for working with R.
- Many more R-related books can be found at https://www.r-project.org/doc/bib/R-books.html.
- For manuals covering the different applications of R see https://cran.r-project.org/manuals.html.

0.6 Reader feedback

Any feedback is greatly appreciated. To provide feedback, please visit our Web page at http://book.ecosens.org and complete the feedback/contact form regarding any typographical errors, mistakes or general ideas on how to improve the book.

1. Introduction and overview

The analysis of geospatial data has been at the core of ecology, geography and other environmental sciences since the very beginning. Answering the fundamental question 'What do we find *where* and *why*?' requires an explicit investigation into spatial relationships, processes and patterns.

Combining geospatial data of different sources, such as field observations and remote sensing data, will enable you to get answers to these questions. This first chapter will get you started with basic geospatial understanding and methods you need to know.

1.1 Spatial data

Spatial, or rather geospatial, data refer to spatially explicit locations. Nearly all the data you sample during field trips are spatial; if you recorded the coordinates of your study site, you could work with them as such.

In principle, a spatial data set consists of two parts: (1) spatial coordinates in a defined coordinate system, such as latitude and longitude; and (2) one or more values, such as a label, a physical measurement or a species observation, associated with this location (Figure 1.1).

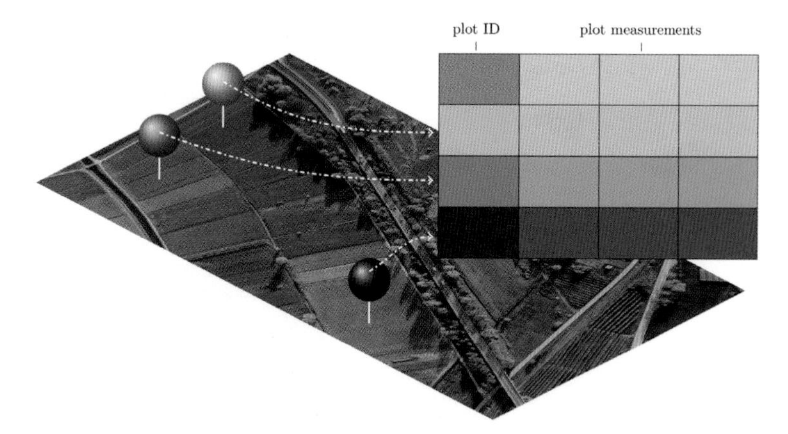

Figure 1.1 Spatial data related to in situ measurement points with coordinates, which locate the data to a specific geographical point in the landscape and optionally to attributes such as a name or a value, for example, the number of species or landcover type. The points are vector data in contrast to the background image (raster data).

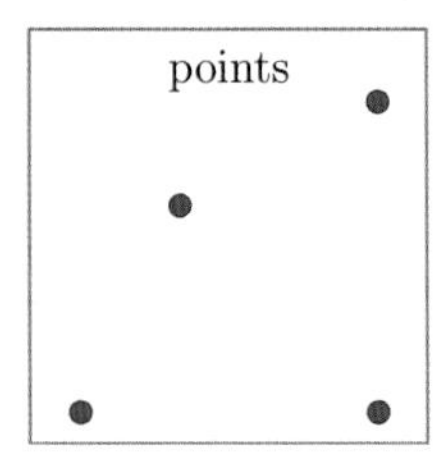

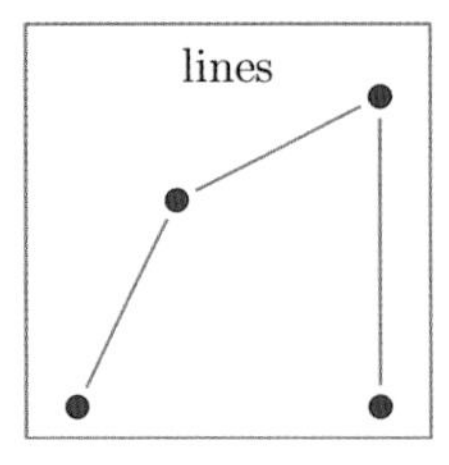

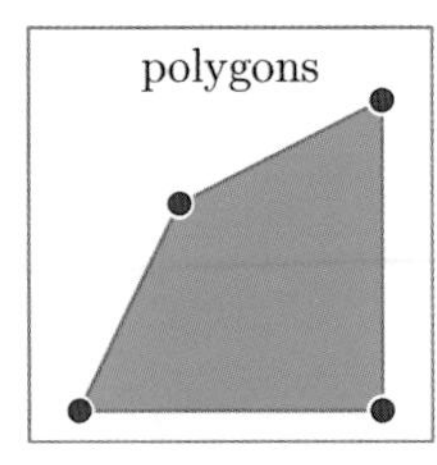

Figure 1.2 Vector data can be points, lines or polygons (areas). Points are coordinates in the geographical space, whereas lines connecting such coordinates and polygons are a closed line between points forming an area. The points in lines and polygons are called vertices and are usually not visible.

Technically you must deal with two types of spatial data, namely **vector** and **raster** data. In general, vector data usually are spatial features like country boundaries, roads or field sites, whereas raster data often represent spectral wavelengths or elevation data. However, raster data can also hold information about distances to study sites and thus provide spatially continuous information of the distance of every cell to a predefined location. In the following sections, we explain these two types of spatial data and apply some essential spatial operations.

Vector data usually represent three types of shapes: points; lines; or polygons (Figure 1.2). The most important difference to raster data is their associated table which, theoretically, can hold as much information as required, ranging from names to values associated with spatial information, such as area or length (Figure 1.3). Raster data, however, can only hold one kind of information within one raster cell. The information can only be numeric; by combining various raster layers, we can hold more than one level

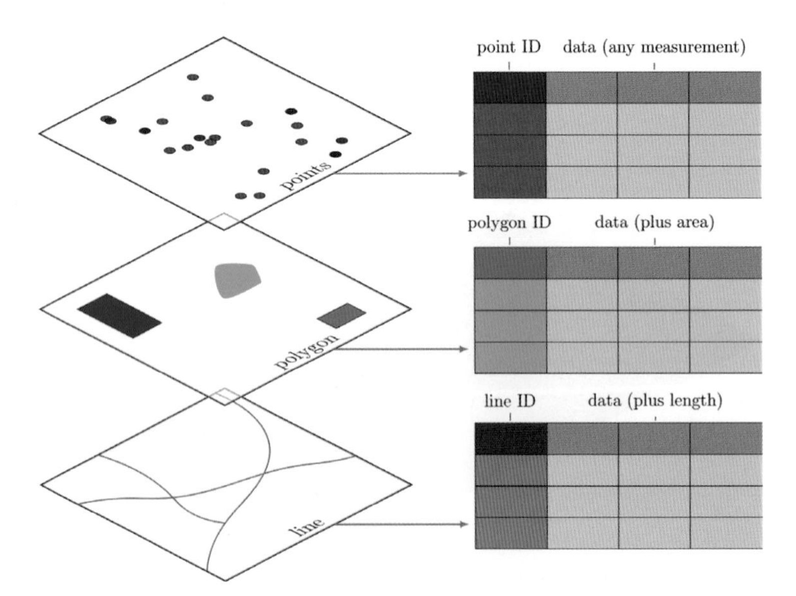

Figure 1.3 Each vector line, point or polygon has a unique ID and can have several unique values, for example, in situ measurement values or names. Additionally, former spatial information can be added in geographic information system (GIS), such as length for lines or area for polygons.

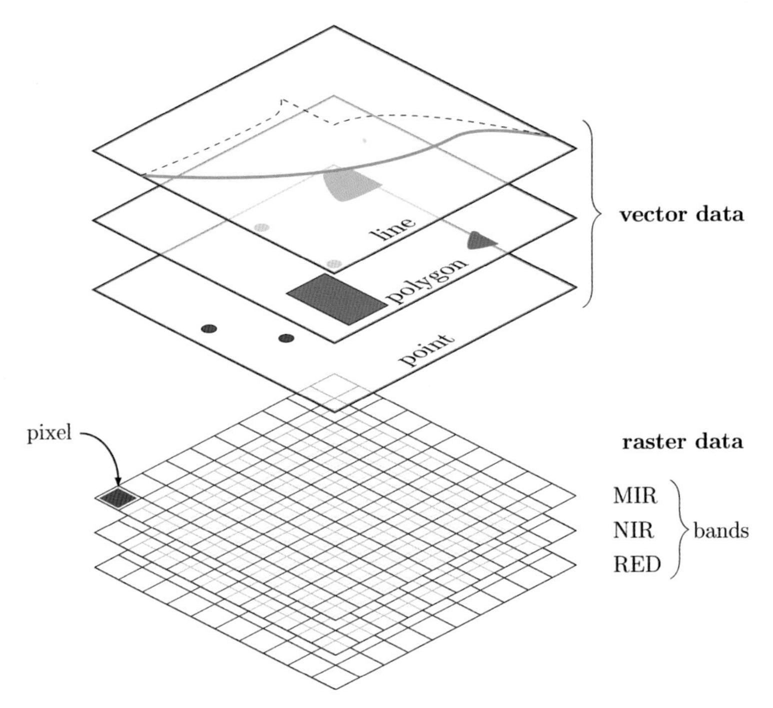

Figure 1.4 Spatial vector data (lines, points and polygons) and different raster bands for one geographical area. All data sets cover the same extent, but they are of different type and hold different information.

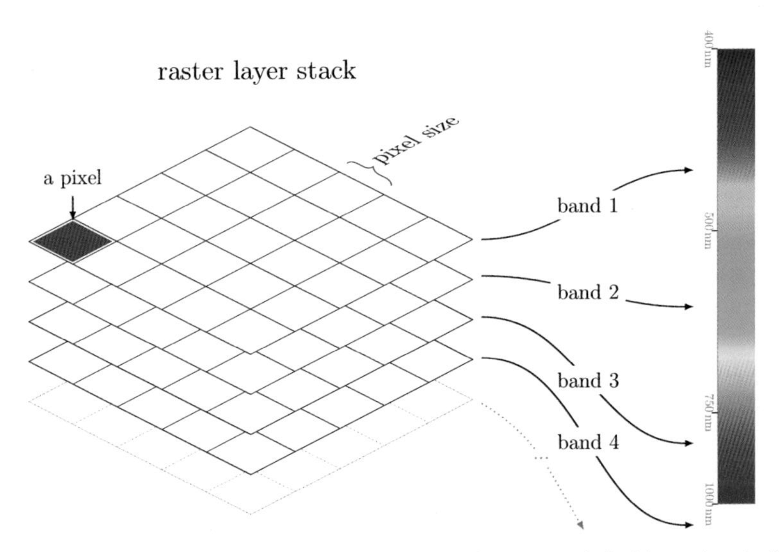

Figure 1.5 Raster data consist of single cells or pixels that can only hold one level of numeric information. Combining different raster bands to a stack allows the merging of various numeric values within one file. The raster bands usually originate from the same sensor acquisition but are not limited to it; further bands, such as elevation (digital elevation model), can be added to the stack.

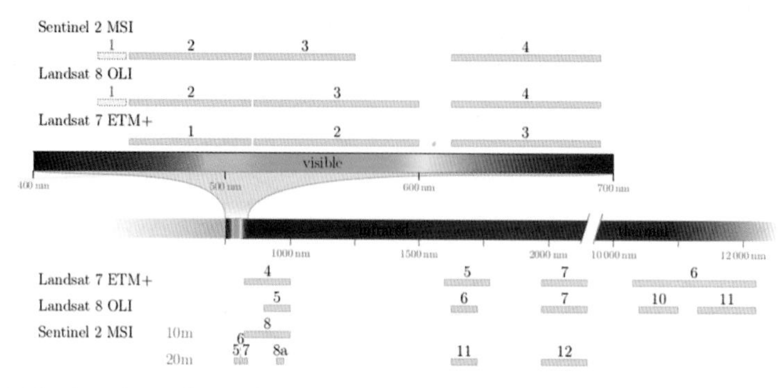

Figure 1.6 The band numbers of three sensors (Landsat 7 ETM+, Landsat 8 OLI and Sentinel-2 MSI) with their position on the spectral wavelength, their width and spatial resolution.

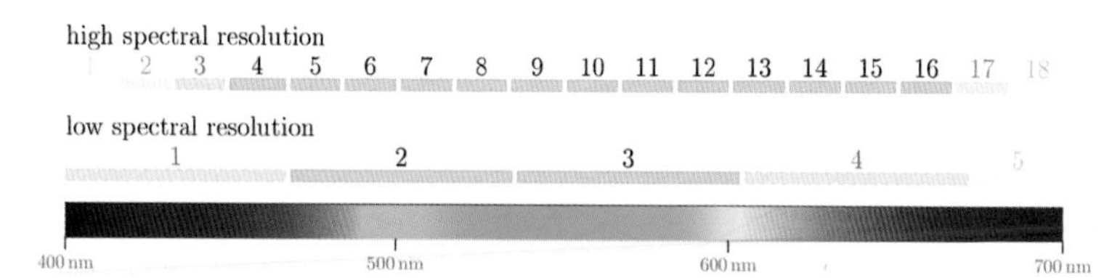

Figure 1.7 The spectral resolution between sensors and data sets may differ. Some have a few broad bands (multispectral) while others provide hundreds of very narrow bands (hyperspectral).

of information in a raster object. These types of spatial data are depicted in Figure 1.4. While they are of different types, these data all cover the same geographical area and can be analysed together, for example, by querying the distance between lines and points or the raster values behind vector points.

Raster data usually represent a certain area and time and often consist of multiple layers. If these layers represent multispectral sensor data, they are also called *bands*. Each band refers to a different wavelength range of the electromagnetic spectrum (Figure 1.5). The order and position of the bands along the spectral wavelength differs between data sets. For example, the sensors operated on the two popular Earth observation satellites Sentinel-2 and Landsat, which are described further on in the book, both provide multispectral land surface information. However, the band number and positions, as well as their width, differ (Figure 1.6). Therefore, it is crucial that you know the origin of the raster data and that you keep such information in the file name (origin, date, processing level, area).

Additionally, remote sensing data sets are differentiated by the number and width of their spectral bands. In the following sections, we use multispectral data sets with a few bands that cover a wide spectral range (Figure 1.7); other sensors provide very narrow bands covering the whole spectrum. In these cases, you may have to deal with hundreds of bands rather than 5–10. Such data are known as hyperspectral data and the information about the number and width of the bands is called *spectral resolution*.

Every single raster band can be plotted in greyscale or in any colour scale and reverted (Figure 1.8). Thus, the colouring of these data sets is totally arbitrary and can be defined by you. The bands of a specific data acquisition are usually combined to process them

Figure 1.8 Raster band colouring is fully user dependent. Data can be displayed in greyscale or using a pseudocolour gradient or its inverted gradient. Different bands may look different due to their cell values, but the colouring itself can be decided by you. The information about the gradient used and its associated values must always be provided.

together and display them as a red-green-blue (RGB) image (Figure 1.9). These colours have nothing to do with the spectral wavelength (red, green and blue); they are just used to colour the single-band values on your screen. Therefore, different band combinations (plotting band 1 in red, band 2 in green and band 3 in blue) can be changed according to your needs. However, to make sure that viewers, for example, colleagues, understand the image colouring, you must always provide information on the RGB colour assignment. Thus, it is common to provide the numbers of the bands used together with the colouring used, for example, 'RGB, bands 4–5–2', indicating that band 4 is displayed in red, band 5 in green and band 2 in blue.

The colouring of three bands (RGB composite) is based on the additive colour system in contrast to the subtractive system used for printing (Figure 1.10). If we display three

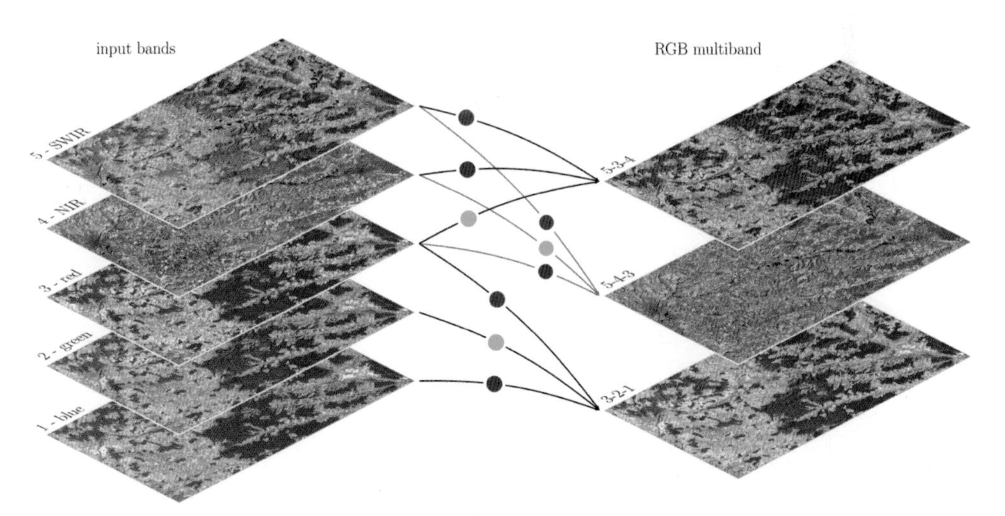

Figure 1.9 Single raster bands are usually combined (stacked) when they originate from the same data acquisition. The combination of single raster data can be displayed alongside individual layers as multibands using red, green and blue colours for three single raster layers; this is known as an RGB plot. Depending on the input data for one of these colours, the resulting RGB plot changes.

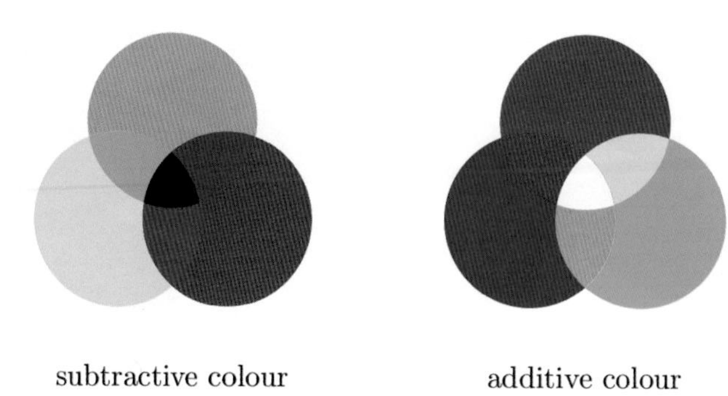

Figure 1.10 Additive and subtractive colour compositing methods used in image colouring and printing, respectively.

bands in red, green and blue and use the bands with the corresponding spectral regions of red, green and blue, we get a so-called natural colour composite because it displays the landscape nearly as we would see it from space.

We have covered certain attributes of raster data, such as spectral resolution. Another important type of resolution is *spatial resolution*, which defines the size of the pixel or grid cells (Figure 1.11). A high spatial resolution, for example, 10 × 10 m would allow us to differentiate more features in the landscape than a coarse spatial resolution (250 m pixel size). High or very high resolutions (1 or 2 m pixel size) is useful to look at because you can see a high level of detail, but it is not necessarily the data you need for your analysis. Should you want to map a forest, but not single tree crowns, it is advisable to use a data set with a 10–30 m spatial resolution. Otherwise, you deal with branches or single trees and not with the forest as a whole.

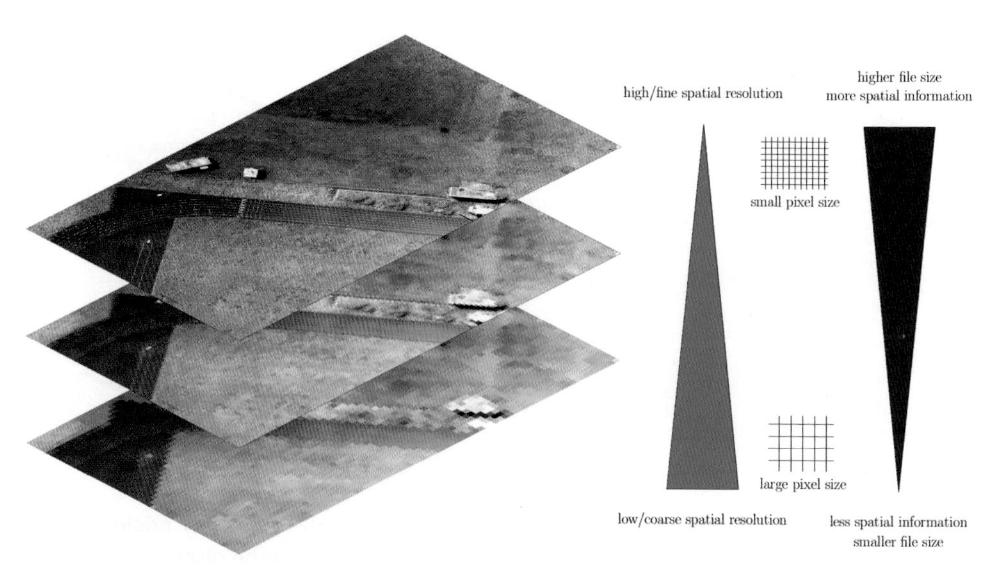

Figure 1.11 Spatial resolution of raster data and the corresponding advantages and disadvantages. A larger pixel size results in a smaller file size and less spatial information and vice versa.

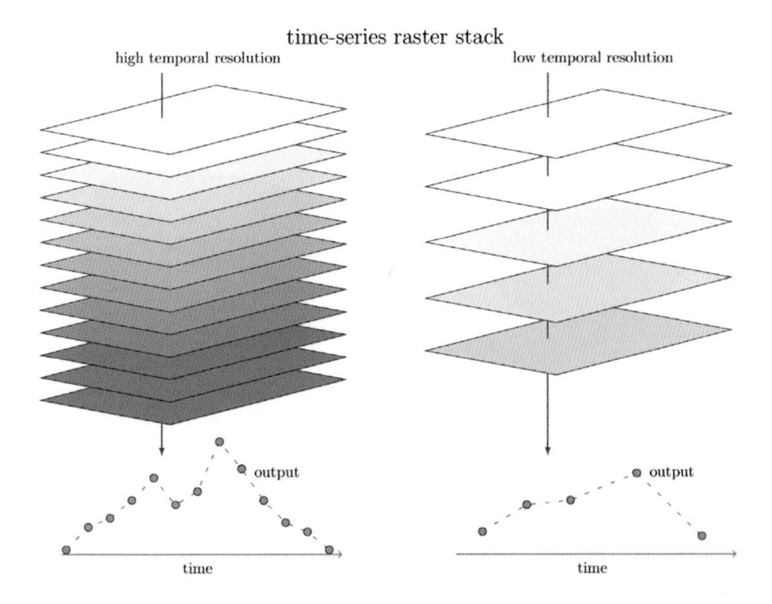

Figure 1.12 Temporal resolution refers to the time interval between two data acquisitions. Remote sensing data can be theoretically acquired every day or once a month depending on the satellite and sensor configuration; however, the actual data availability might differ due to technical or environmental limitations.

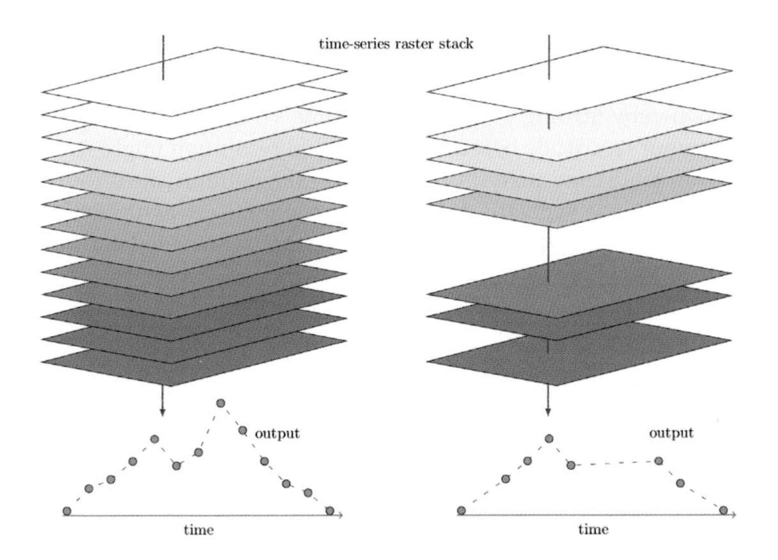

Figure 1.13 The actual availability of a theoretically regular time series can deviate significantly. Irregular availability of environmental data is rather common. Very often, an actual temporal raster stack has missing data due to no available imagery or cloud cover.

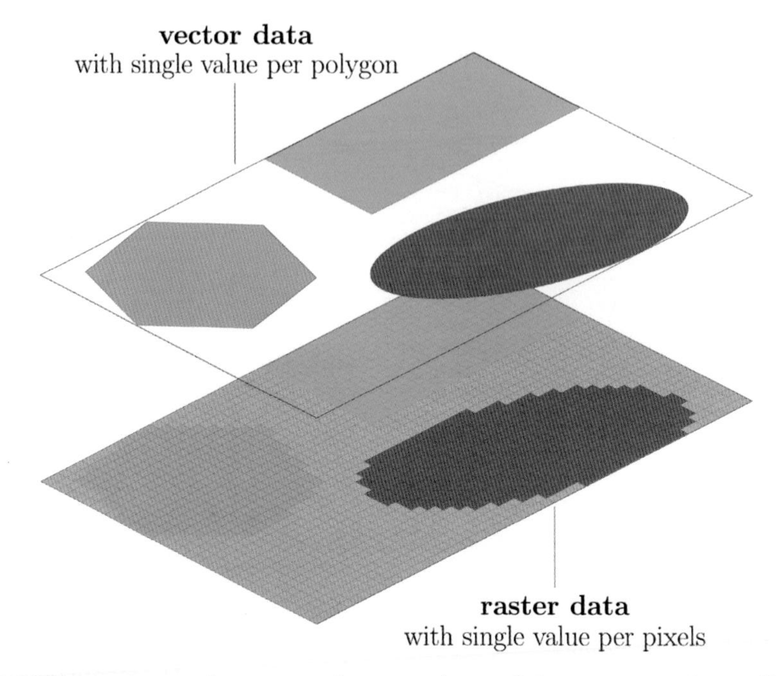

Figure 1.14 Vector data can be converted to raster data and vice versa to apply specific analysis only available for a specific spatial data type. Depending on the chosen spatial resolution, a vector polygon might look smooth or coarse after being converted to a raster. The colouring of vector and raster values can be defined by you and requires a notation to know the value associated with each colour.

Another key type of resolution is *temporal resolution*. It refers to the amount of (theoretically) available images of the same location over time: many acquisitions over time provide a high temporal resolution and vice versa (Figure 1.12). Theoretically because a single data set might not be available, for example, due to cloud cover or other obstacles, thereby leading to gaps in your time series (Figure 1.13). This is crucial when analysing your landscape because the temporal resolution might cause significant differences in your landscape analysis. If the temporal repetition rate is too coarse to map a short-term peak in your vegetation signal, you might falsely assume that the peak of the vegetation signal is earlier in the year than it is (compare the left and right graphs in Figure 1.13).

You can also transform between raster and vector data (Figure 1.14). This is called **rasterization** or **vectorization** and comes with certain provisos. For example, a raster is of a predefined cell size whereas a vector has no specific spatial resolution; thus, when a vector is converted to a raster, the spatial resolution must be defined. Specific issues with such conversions and how to deal with them are addressed in the following chapters.

Such spatial data are managed in a GIS, which also provides a wide range of functions to achieve many tasks. However, in specific cases, you might need to apply other software packages or custom-build certain functions yourself. Having said that, all tasks outlined in this book can be achieved using QGIS.

1.2 First spatial data analysis

The topics outlined in the previous section are best learnt by applying them. We will conduct an initial analysis using an open-source GIS. Some common tasks will be executed on an artificial example to outline the approach; however, we use an actual GIS that we will also use for the actual data analysis in subsequent chapters.

In the following sections, artificial data are used to outline the general approach using simple spatial objects. The data sets are in the folder 'data_chapter 1' and consist of raster (grd) and vector (GPKG) files. The raster files consist of several single rasters ('b1. grd, b2.grd ...'), a multiband raster ('stack_b1b4.grd') and a classified raster with five categorical values representing land cover classes ('unsup_class.grd'). These data sets originated from real Sentinel-2 data and are used to create an unsupervised classification with five classes. More details on the sources of these data and required operations are provided in the next chapters. The vector data are point, line and polygon objects with random values in the corresponding table.

The data can be downloaded from http://book.ecosens.org and should be stored in a new folder. We recommend to (1) save all spatial data in one place for one specific analysis (in our case, the chapters of this book) and (2) use subfolders for different data types, for example, vector and raster files.

1.2.1 A GIS

QGIS is a very user-friendly GIS available for free from https://www.qgis.org/en/site/. It provides an appealing graphical user interface (GUI) and is a very intuitive software package. It is used throughout the book and is the ideal GIS to get you started with spatial analysis. The QGIS interface (Figure 1.15) is especially helpful for interactive data analysis, such as data display, digitization or map generation. The map view window displays the data, while the layers are listed on the left and all the toolbars for specific operations are located at the top of the interface.

QGIS has a very high development pace: new releases or beta versions are announced every couple of months. Usually, three QGIS versions are released per year. Odd-numbered versions are development versions while even-numbered versions indicate stable releases. This book is written using the most recent development version to provide you with the most up-to-date QGIS version.

QGIS or GIS in general provide a substantial amount of spatial data analysis functionality that is hard to cover within one book. For QGIS, there is an extensive online

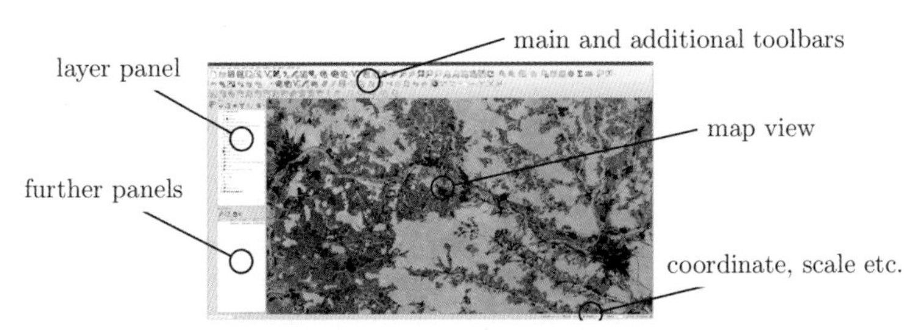

Figure 1.15 QGIS interface showing the layer panel on the left and several analysis buttons on the top and left.

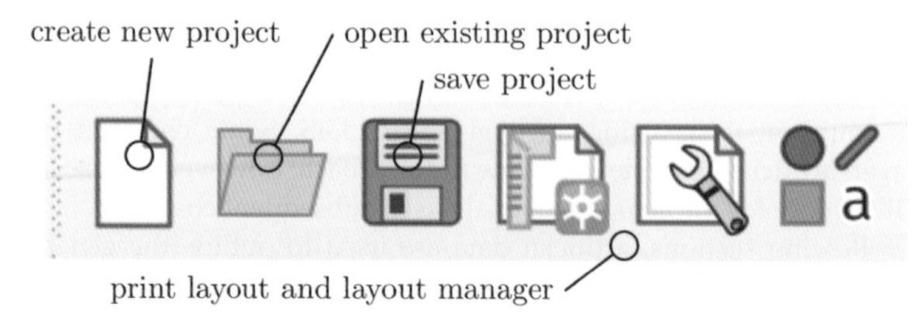

Figure 1.16 Left to right: project toolbar with buttons to start a new project, open an existing project, save a project or create a new map project.

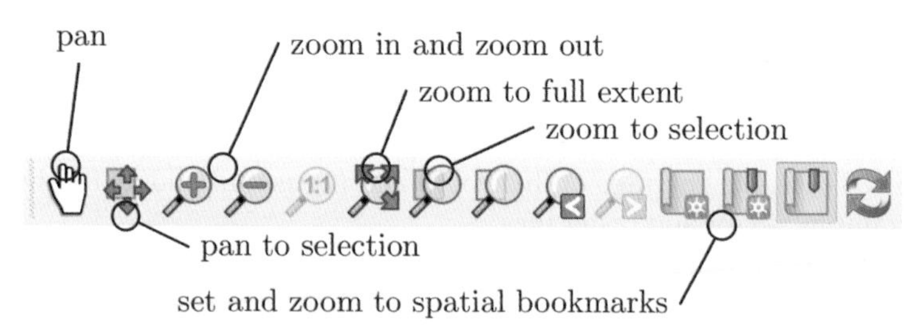

Figure 1.17 The navigation panel provides functions to navigate through the map, such as zooming in or out. Moreover, spatial bookmarks can be set to navigate easily to a predefined area.

archive of help pages and tutorials. Go to *Help > Help Contents* or hit F1 in QGIS and your browser will be directed to the help pages of your current version of QGIS.

You will quickly realise that QGIS is more than a simple GIS. It has fully fledged support for high-end geospatial data processing with all the functionality you would expect from a software for spatial science. While you can achieve everything in the GUI, QGIS also provides a graphical model builder to generate program routines for users not familiar with a scripting language. More importantly, it also includes a scripting console for the Python language. Furthermore, QGIS provides interfaces to R, GRASS, SAGA and other software programs so you can use their additional functionality within QGIS.

QGIS comes with various default panels and toolbars that serve specific purposes and are visible after opening QGIS, such as the *project toolbar*. This toolbar provides options to create a new project, save it or open a print layout (Figure 1.16).

The navigation toolbar allows you to move through the map, zoom in and out and create spatial bookmarks to easily zoom back to a saved area (Figure 1.17). The icons are quite intuitive; if you hover the mouse above them, a short explanation is shown. The most important ones for the moment are the first three from the right: *pan*; *zoom in*; and *zoom out*.

The layer toolbar is the layer that needs to be used first because it opens an interface to load spatial data into your GIS (Figure 1.18). Vector, raster or delimited text layers (first three buttons from the left) can be loaded through this toolbar. Other formats can also be loaded using the other buttons, which will open the same *Data Source Manager* window (Figure 1.19). Within this manager, you can browse to the data set you would like to load and add it to your project.

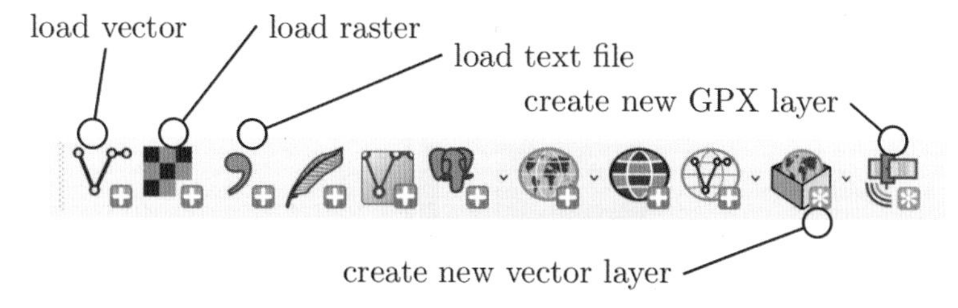

Figure 1.18 QGIS interface showing the layer panel on the right and several analysis buttons on the left.

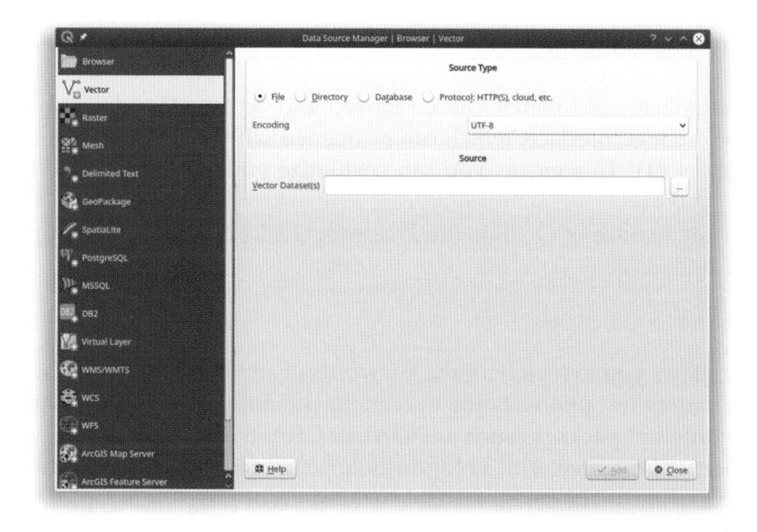

Figure 1.19 The QGIS Data Source Manager allows the user to import any file into QGIS, such as vector, raster and delimited text files (field site coordinates). Alternatively, locally stored data can be loaded by dragging and dropping it into QGIS from the file manager.

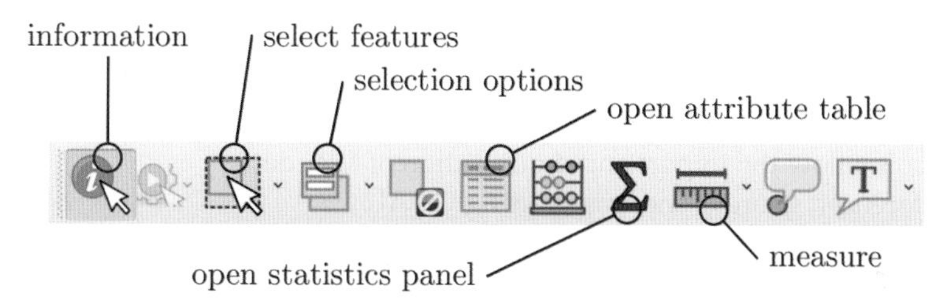

Figure 1.20 The Attributes toolbar provides information about your spatial object, such as the table associated with the vector file.

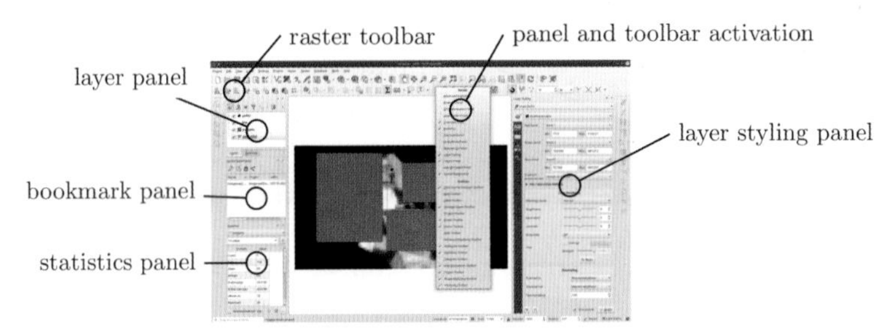

Figure 1.21 The QGIS layer/toolbar activation interface, showing various activated panels and toolbars for individual customization.

The *Attributes toolbar* allows you to query the spatial features in your view or select parts. Additionally, it provides a button to open the corresponding table of a vector file (Figure 1.20).

All these toolbars and panels can be moved and activated or deactivated through the toolbar and panel interface, which can be opened with a right-click on any panel or toolbar (Figure 1.21). This means you can customize your QGIS interface to suit your needs.

Another key feature of QGIS is the *Processing Toolbox*, which can be opened using Ctrl + Alt + T or:

```
Processing > Toolbox
```

This toolbox allows you to search for the keywords of any operation you want to execute and offers a variety of options. Some operations are native QGIS functions, others are provided by different packages such as GDAL or GRASS (Figure 1.22).

In addition, QGIS comes with a plug-in architecture for which users have contributed a variety of extensions. The set of available extensions or plug-ins is rapidly evolving as is QGIS. QGIS comes with a plug-in manager that makes it easy to browse and install plug-ins from a central repository.

Some recommended plug-ins you might find useful are:

- OpenLayersPlugin or QuickMapServices: displays Google Earth, OpenStreetMap, Bing Maps, etc. with imagery in the background.
- MMQgis: vector analysis and management.
- Quick Finder: searches the Web or your local layers.
- MapSwipe Tool: allows to swipe between two layers.
- Profile tool: creates profile lines.
- GPS tools: allow the import and modification of GPS data.
- Georeferencer: georeference raster using GDAL.
- Processing toolbox: interfaces to GRASS, Orfeo Toolbox, R and SAGA.

These and many more can be browsed and searched for at http://plugins.qgis.org/plugins/ and installed via:

```
Plugins > Manage and Install Plugins ...
```

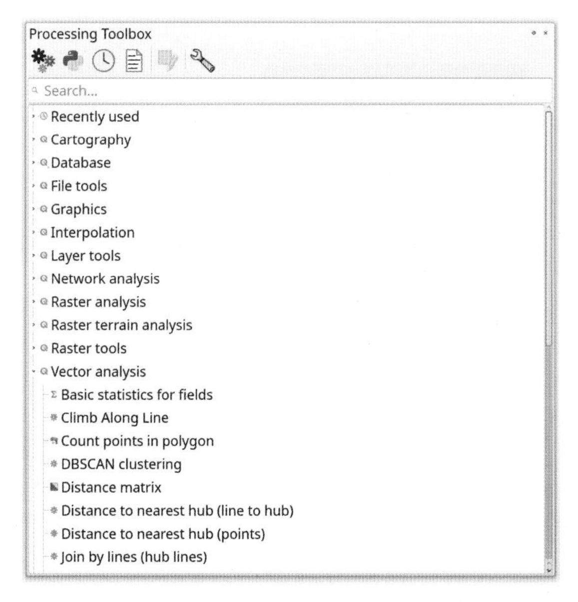

Figure 1.22 QGIS processing toolbox interface showing further geospatial operations accessible within QGIS. Operations can be searched and recently used operations are listed separately.

> **QGIS INFO**
>
> QGIS functionality can be increased by using SAGA, GRASS or Orfeo Toolbox functions. The new functionality will then be available through the processing toolbox in QGIS and no need to learn another software program.

Vector data

We first load our artificial vector data into QGIS using the *vector* button, which opens the *Data Source Manager* for vector files. Here, we browse and select the GPKG files (point, line and polygon). After doing so, you should have all three spatial objects in your *Layers Panel* on the left and displayed on your screen (Figure 1.23).

We can move the order of your layers in the *Layers Panel* using drag and drop with the left mouse button. We can also (de)activate their visibility by checking the square to the left of the layer name. More options are available for each specific layer when doing the following:

```
Right-click on layer name >
```

The options include:

- Zoom to layer: this changes the view window on the right to the extent of the present layer.
- Zoom to selection: if you selected a specific point, polygon or line, the view would change to this specific extent.

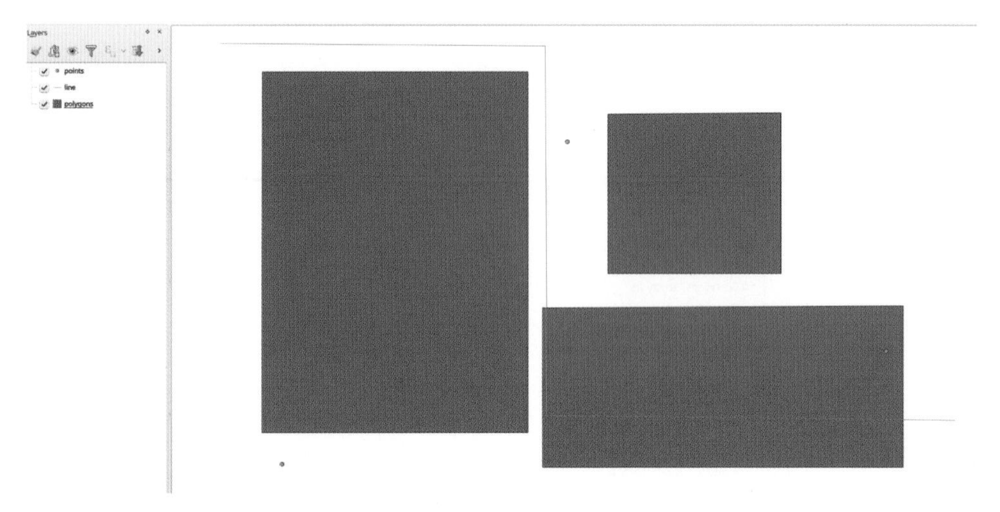

Figure 1.23 QGIS interface showing the layer panel on the left and various analysis buttons on the top and left.

- Show in overview: a great feature to keep the overview. If you select it and add the *overview panel*, the layer will be displayed in an overview. Thus, you can always see where you are and move the rectangle in the overview, which changes the view window.
- Show feature count: it displays how many features are in your vector object.
- Copy/rename/duplicate/remove layer are self-explanatory.
- Filter: this is a great way to filter your spatial data by values within your table.
- Set layer scale visibility: it allows you to display your data on certain scales; for example, if you zoom out, they will eventually stop being displayed.
- Set CRS: this is an important feature because it allows you to set the projection for your whole project (set project coordinate reference system (CRS) from layer).
- Export: saves your layer to your hard drive; usually important for newly created files or if you want to save them in a different file format.
- Styles: very useful feature to copy and paste styles across different vector objects.
- Properties: displays all the actual properties of your vector file (Figure 1.24).

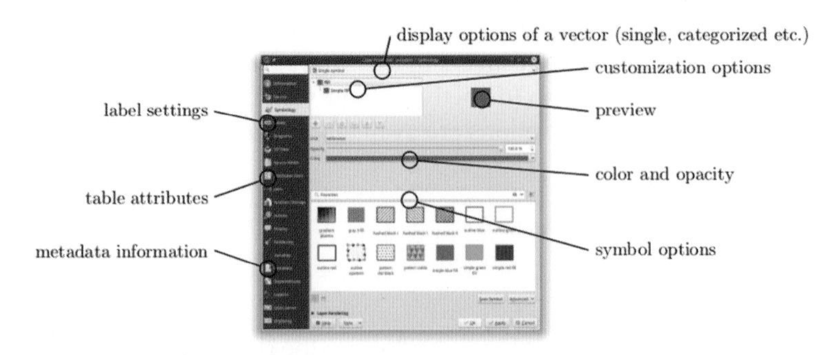

Figure 1.24 The QGIS interface for vector properties provides various customization options and detailed information about this spatial object as well as certain operations such as joins (see Part 2).

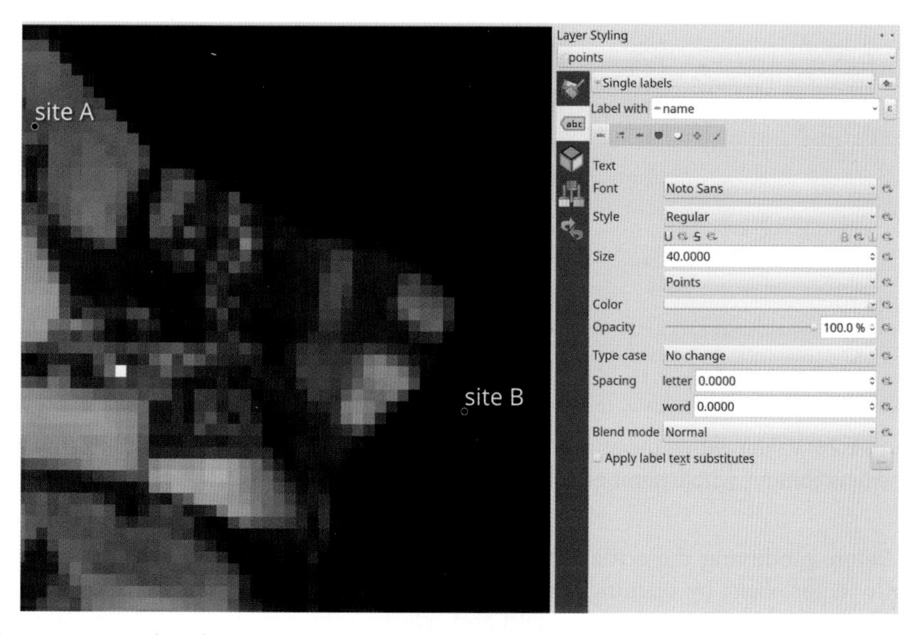

Figure 1.25 Within the vector Layer Styling panel, the look of the vector as well as its labels can be defined. Depending on the background, a drop shadow or different text colour might be more appropriate.

The *Properties* interface provides various relevant details about your object as well as options on how to modify its appearance. The colour or symbol can be changed either for all vector features or categorized according to a value in the table. At the top of the window (Figure 1.24), display options are available via a drop-down menu. *Single symbol* is the default, which means that all features are displayed with the same colour or shape. Changing it to *categorized* or *graduated* allows you to colour the spatial features differently using categories or value gradients. Further information and options are available via the tabs on the left, such as *Metadata* or *Labels*.

To style layers, another panel can be used where the same vector style modifications can be applied. *Right-clicking* on any panel and selecting *Layer Styling* will activate this panel. The colour and shape are defined in the first tab while labels can be activated, and their layout can be changed in the next tab. After setting *Single labels* and which column should be used for the label, we can define the size, colour or *drop shadows* for the text (Figure 1.25) as well as the position or background. Selecting another layer in the layer panel will change the *Layer Styling* panel accordingly. The layer to be styled can also be set within *Layer Styling* through the drop-down menu on the top.

TASK

(1) Define your own colour ramp by opening the *Select Color Ramp* interface (click on the colour ramp button) and adding new colour stops and gradients.
(2) Set point sizes to be related to a column value within the *Layer Styling* panel by selecting *Graduated* and *Method = size*.
(3) Filter and create new columns in the attribute table interface.

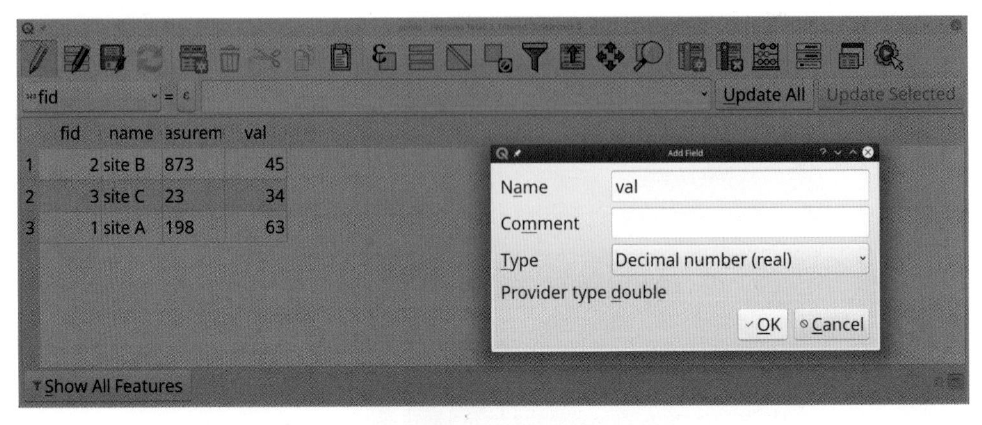

Figure 1.26 Values within the attribute tables can be edited or new columns added. The type of the new column, whole number or binary, need to be defined on creation.

Before modifying the actual data, we can already change the corresponding table of our vector object by opening the table (select icon on the *Attributes toolbar* or right-click on layer then *Open Attribute Table*) and selecting the *pen* icon, which activates the editing (Figure 1.26). After the editing is activated, new columns or rows can be added (icons with a star) or removed (icons with a cross). Additionally, new columns can be created using the *Field Calculator* (slide rule icon). Further options are also available, such as showing the table below the map view or filtering values within the table.

Spatial data modification

A GIS is mainly a tool for modifying spatial data. The number of spatial data modification options are numerous; we will introduce and demonstrate a few of them in the following sections. How to create your own vector data will also be introduced in Part II.

Common tasks in spatial data analysis are:

- Buffering: creating an area around an existing location to analyse its surrounding.
- Clip: cutting one vector file based on another.
- Difference: creating spatial information where two areas are (not) overlapping.
- Intersection: checking which study areas intersect with another area, for example, a buffer around roads.
- Union: combine different adjacent or overlapping vector objects and make them one because you may not be interested in single buildings but the overall urban area for later spatial analysis, such as buffering and intersection.
- Distance analysis, spatial statistics and further options.

The general approach is outlined in Figure 1.27, where the different spatial operations and the corresponding results are visualized. Operations can also be based on one another as the combination of *union* and *dissolve* shows (Figure 1.27). However, also the input order is crucial to consider as the same operation with the same spatial objects, but as different inputs will result in fundamentally different results (Figure 1.28).

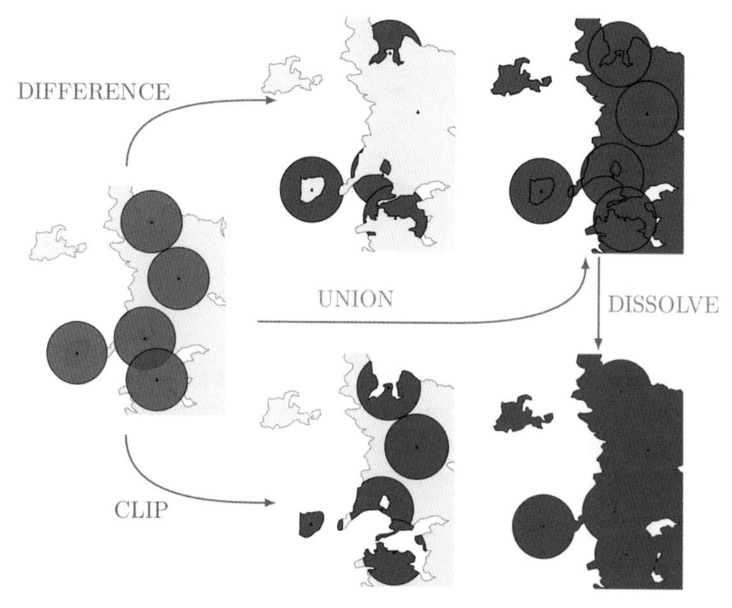

Figure 1.27 Resulting spatial objects (green) with the original input in grey for different spatial operations and consecutive spatial operations and their output.

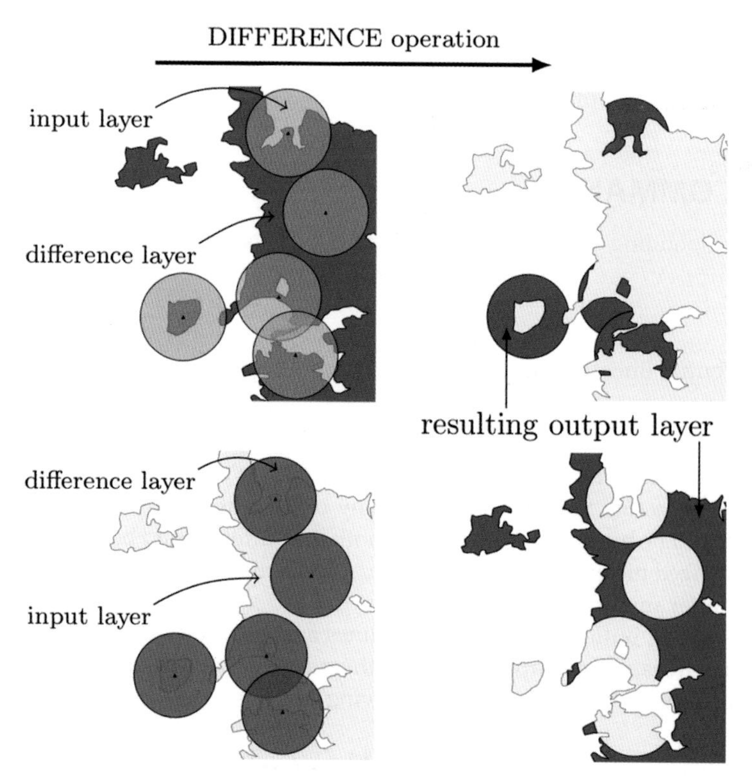

Figure 1.28 Different results of the difference operations caused by assigning the input layers differently.

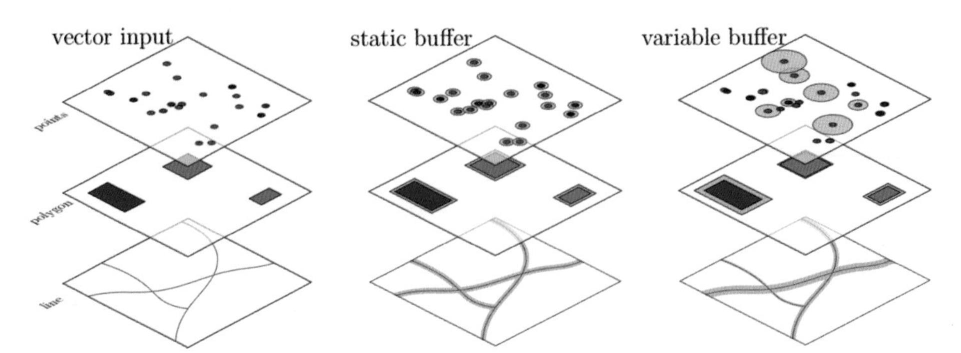

Figure 1.29 Any vector file can be buffered based on values for all objects or a varying value within the corresponding table of the vector object. Variable buffer sizes could be necessary if a factor, for example, the intensity of traffic or noise along a road, is relevant for your analysis and has a larger impact on the surrounding area than a foot path.

Buffering

A very common task is the buffering of points, lines or polygons. This means that, based on a predefined distance or an attribute in the vector table, a new area is created around this feature. Any spatial object can be used to create a buffer around it. The size or distance from the origin needs to be defined and is usually set as metres, feet or kilometres in case of projected data. The buffer size can also be extracted from a column in the corresponding vector table and result in different buffer sizes for single spatial objects (Figure 1.29). Polygons can also be buffered into the area (inside the polygon) like the common outside buffering of the shape.

> **R COMMANDS**
>
> The corresponding command in R provided by the *sf* package for a buffer of 100 m is: `st_buffer(vector_object_name, 100)`

If you buffer lines, points or polygons, the same operation applies:

```
Vector > Geoprocessing Tools > Buffer ...
```

Within the buffer menu (Figure 1.30), various settings can be defined, such as: (1) the distance of the actual buffer in metres or any other unit; (2) whether only the selected features should be buffered; or (3) how smooth the buffer should be (*segments*). The latter can be visualized nicely if applied to a circle, since increasing the *segment* value makes the resulting buffer smoother.

The resulting buffering of the points, lines and polygon objects is shown in Figure 1.31. Certain areas are overlapping, while others are intersecting one another. This spatial information is further explored in the following sections where further spatial operations are applied.

Buffered files are saved as temporary files by default, unless you use the interface to define an output file. Saving a temporary file to your file system can be achieved as shown in Figure 1.32:

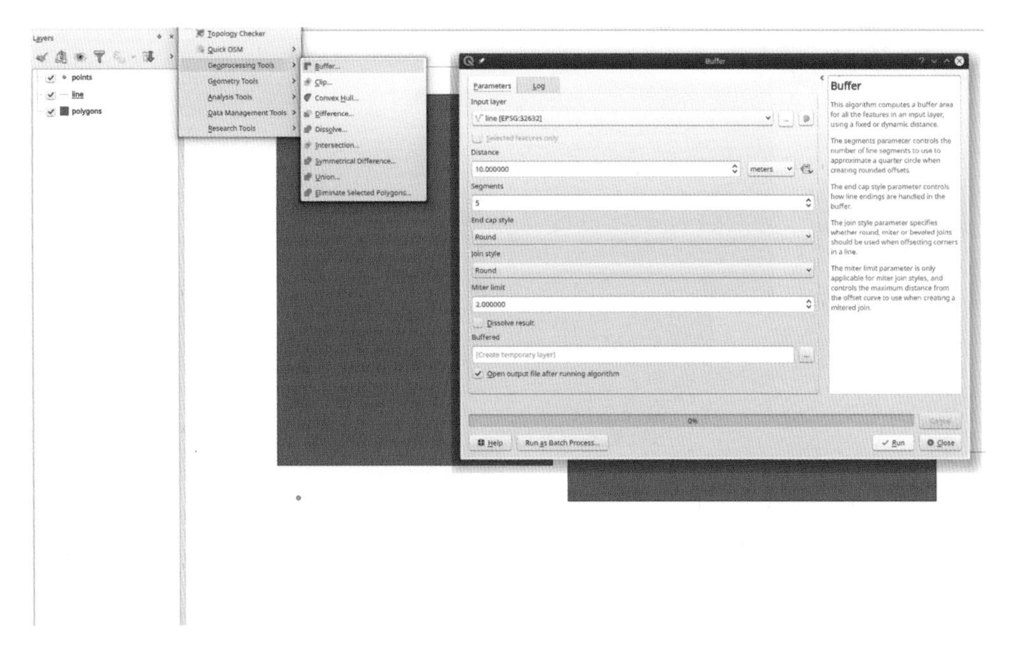

Figure 1.30 Within the buffer menu, the distance of the buffer can be set using different units. Additionally, smoothness can be defined by segments if only selected features are buffered and if the buffers are to be dissolved.

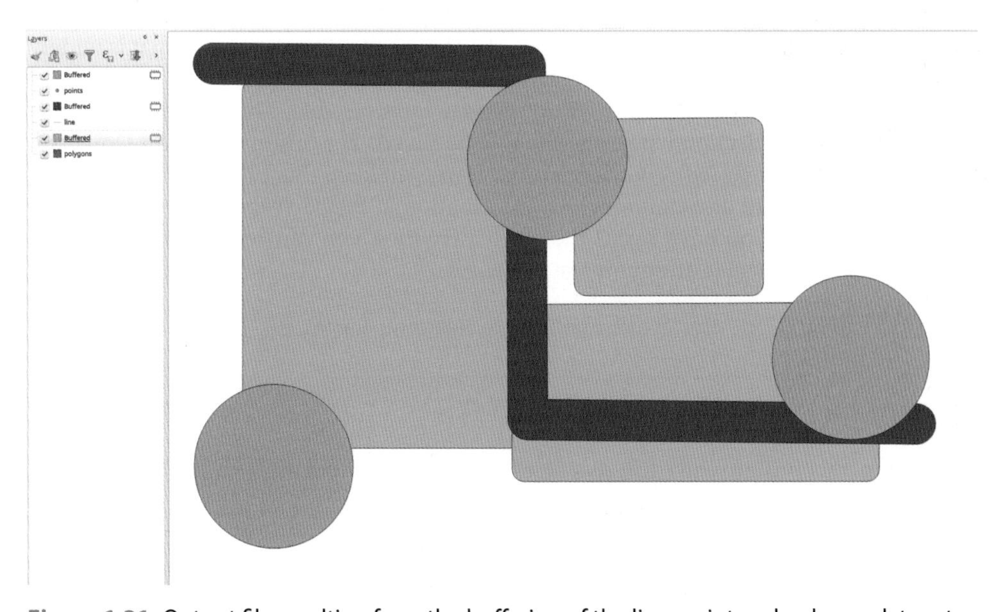

Figure 1.31 Output file resulting from the buffering of the line, point and polygon data set.

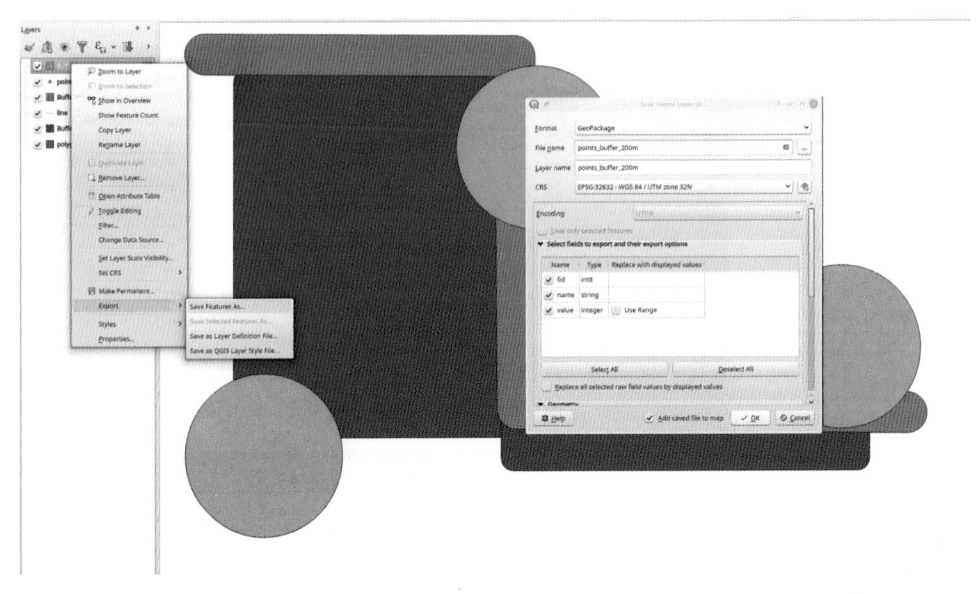

Figure 1.32 When a new file is saved as a temporary file, it must be saved before QGIS is closed. Within the menu, the projection as well as the table entries are displayed.

```
Right-click on the layer name > Export > Save Feature As ...
```

Within the interface, you must provide the name of the file and the location you want to save the file to, as well as the file format. We have used the *GeoPackage* (.gpkg), rather than the *shapefile* (.shp) format because it consists of only one file. The *shapefile* format is widely used but comes with a variety of corresponding files; all these files need to be copied otherwise the whole spatial object will no longer be accessible. The CRS also needs to be set. This is the projection where the file will be saved. The original file is in Universal Transverse Mercator (UTM) WGS84 zone 32N; we will keep this projection for the new vector files. You could also define which information in the corresponding table is kept (Figure 1.32). After you have saved the temporary files, you can remove them because they will not be accessible after you restart QGIS:

```
Right-click on the layer name > Remove Layer
```

TASK

Create further buffers with different sizes and modify the segment value. What does *segment* do?

Clip, intersect, dissolve and union

There are numerous options for spatial operations. As an example, we will apply *clip* or *intersect, dissolve* and *union* to outline the general idea. Depending on your research needs, different operations or a series of consecutive operations are needed to obtain the desired output.

Clip and *intersect* are quite similar but differ in the way they deal with attributes. *Clip* keeps the attributes of the primary input not the object used for clipping, while intersect keeps the attributes of both objects. Both represent a spatial query of which areas of two objects are overlapping. The output is a spatial object that represents the area covered by both input objects (Figure 1.33). Such analysis is usually applied if you need to know which parts of your study area are overlapping with the buffer of a road and might be impacted by it.

> **TASK**
>
> Explore spatial queries to check which features overlap or intersect.

In contrast, *Union* combines different spatial objects but keeps the information of the shape. Thus, you can still see the lines of the features, but all are now one spatial object. However, every single section can now be queried and modified. By applying *dissolve*, you remove these segments and the different spatial parts will be one large object without any subdivisions.

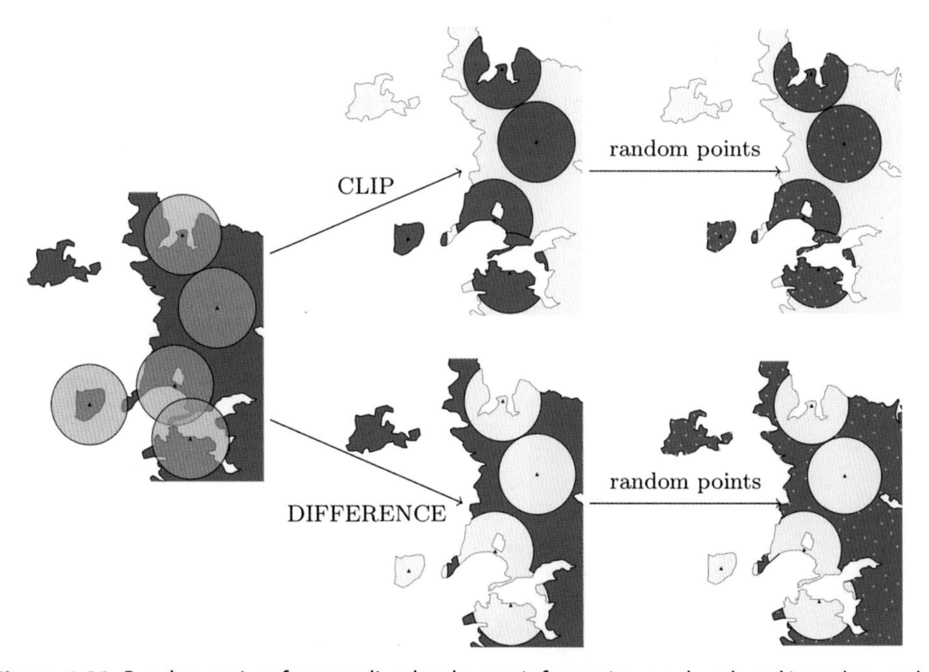

Figure 1.33 Random points for sampling land cover information can be placed into the newly created spatial subsets. Based on the settings, different areas are used to place the random points.

R COMMANDS

The corresponding commands in R provided by the *sf* package for a spatial manipulation are:

```
st_join()
st_intersects()
st_within()
geom_combine()
```

In QGIS, a variety of vector operations are available. Common operations, such as *clip* or *intersect*, can be accessed via:

```
Vector > Geoprocessing Tools > Difference ...
Vector > Geoprocessing Tools > Clip ...
Vector > Geoprocessing Tools > Union ...
```

The *difference* window and the resulting output are shown in Figure 1.34. Using the polygon as the *input layer* and the point buffer as the *overlay layer* results in a new vector object that has the shape of the polygon, but with all areas covered by the point buffer cut out. Such operations may be relevant if new study sites should be selected that are not within a certain distance of pre-existing in situ measurements.

TASK

Try the difference, clip, union or other operations on the different vector objects. Also swap the input and overlay layer and explore the resulting spatial shapes. Think about research tasks where such operations might be useful.

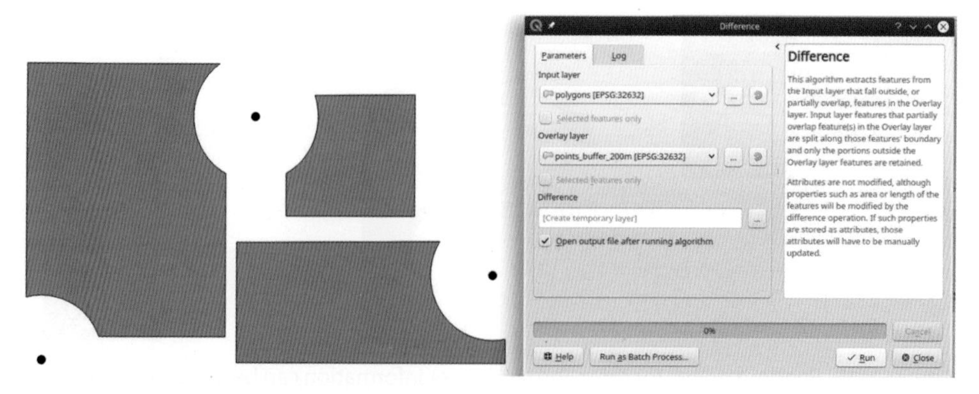

Figure 1.34 Difference operations using the polygon and point buffer objects result in cut out areas within the polygon based on the shape of the buffer areas.

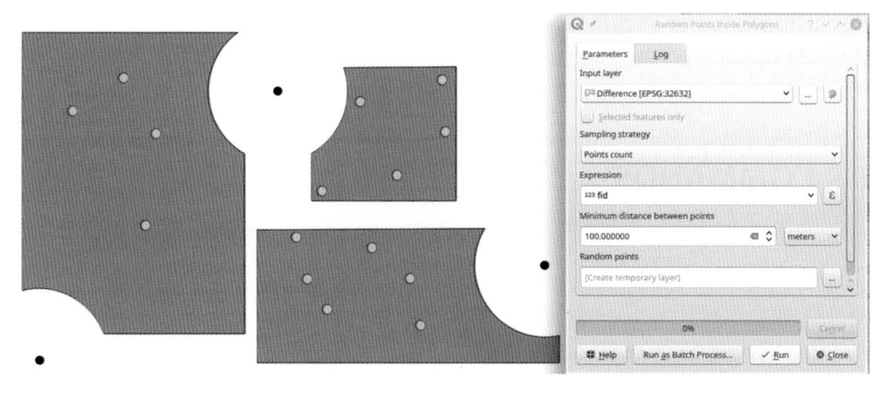

Figure 1.35 Random points can be placed into the newly created spatial subsets. Based on the settings, different methods are used to place the random points. Here the number of random points is taken from a specific column and the minimum distance between points has been set to 100.

> **QGIS INFO**
>
> In the layer legend on the right-hand side, temporary objects have a 'memory' icon; clicking on it will open an interface to save this specific object.

Adding new objects and spatial information

The modified spatial objects could also be used for a variety of further analysis, such as spatial sampling in selected areas only. This is done by placing random points in the newly created spatial polygons (Figure 1.35) via:

```
Vector > Research tools > Random Points Inside Polygons ...
```

The column *fid* is used to define the number of random points per polygon; using other columns with different numbers will result in a different output. Moreover, the minimum distance between random points can be set if points must be apart by a predefined distance.

Additionally, we can add geographical information to the tables of our vector, such as length, area or perimeter:

```
Vector > Geometry Tools > Add Geometry Attributes ...
```

Applying this operation on the polygon object and the *difference* output shows the divergent spatial attributes due to the different shape and area.

Raster data

Raster data usually has file sizes larger than you need; hence, you may want to reduce the geographical extent of these files. Additionally, you might want to mask out certain areas of no interest because they are not located within your study area. Such tasks are usually done by using existing vector files. In some cases, you may need to convert a raster to a vector file or vice versa (*Polygonize* or *Rasterize* in QGIS within the *Raster conversion* menu).

> **🛈 R COMMANDS**
>
> Commands related to vector <-> raster conversion in R are provided by the *raster* package:
>
> ```
> rasterize()
> rasterToPoints()
> rasterToPolygons()
> ```

Raster data import

At the beginning of a QGIS session or any analysis, we first must import the raster data into our GIS. Some data sets come as single-band rasters, for example, an existing land cover classification; others come as single files but belong together. Both types need to be loaded first into QGIS via the *Add Raster Layer* button or

```
Layer > Add Layer > Add Raster Layer ...
```

Alternatively, you can also drag and drop the files from your file system in the *Layers Panel* or map area.

> **🛈 QGIS INFO**
>
> Many tasks, such as adding a raster layer, can also be achieved using certain keyboard combinations, for example, CTRL+SHIFT+R to add a raster layer and CTRL+SHIFT+V to add a vector layer.

We added the following raster data sets: unsup_class.grd (discrete values of five classes); and an example data set of four bands of spectral reflectance information belonging together: b1.grd, b2.grd, b3.grd and b4.grd, with

```
Layer > Add Layer > Add Raster Layer ...
(alternatively, using the button in the Manage Layers Toolbar
or by dragging and dropping into QGIS)
```

All these bands will show up at the top of your layer panel. If they cover another object you want to have on top, just drag and drop the layer where you would like it to be. The layer panel is represented as a pile of paper. The top layer will cover the layers below unless we have a vector file covering just parts of the area or a raster file with values being masked out. Alternatively, you can click on the small square left of the raster name in the *Layers Panel* to activate or deactivate it.

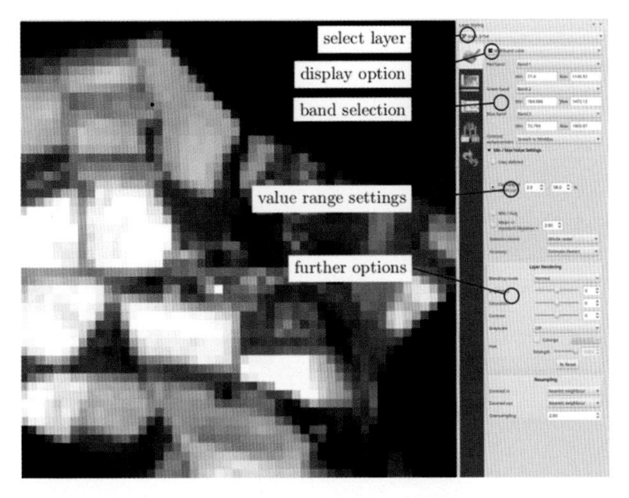

Figure 1.36 Raster objects are usually displayed in greyscale and can be changed to a single-band pseudocolour gradient. In single bands, band selection can be ignored but using a multilayer raster, single bands can be displayed via this drop-down menu.

QGIS INFO

Any spatial data can also be added to QGIS by dragging and dropping it from your file manager.

The raster layer is displayed in greyscale by default but can be changed to any colour scale (see Figure 1.8). Within QGIS, the colour can be changed via the *Layer Styling* panel or *Symbology* tab within the raster *Properties* interface. Changing *Singleband gray* to *Singleband pseudocolor* allows to assign any colour gradient to our raster (Figure 1.36).

TASK

Change the colour gradient of your single raster, change the input layer and the display options. Check what happens to the displayed raster.

Usually, and as explained previously, several bands belonging to one acquisition are downloaded and saved as single files even though they belong together. This is not a necessity, since many operations can be achieved using single files. However, storing all layers that belong together in one file is advantageous. In the following, the loaded raster objects (b1–b4) are combined (stacked or merged) with:

```
Raster > Miscellaneous > Merge ...
```

As shown in Figure 1.37, the single raster layers that are loaded beforehand are listed and can be added to the layer stack. Files can be dragged inside the *selection* window to the top or bottom to maintain the correct order, which usually corresponds to the band number. If the order deviates from a commonly used order or if only some of the bands

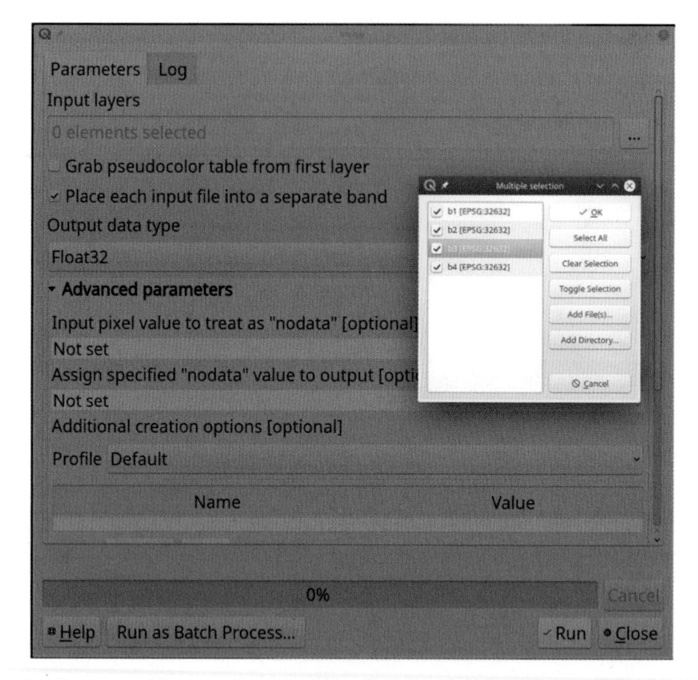

Figure 1.37 Single raster files can be combined into a raster stack to keep files that belong to the same acquisition together. This also allows to plot the raster layers as an RGB colour composite.

are selected, the information needs to be provided within the new raster stack. After completing the merge operation, the single raster layers can be removed from the layer panel. In subsequent steps, only the new layer stack needs to be loaded.

After merging the single layer, you can now display the data as an RGB plot (see Figure 1.9) using the Layer Styling interface or the *Symbology* tab in the *Properties* interface. Within this interface, you can define which band values should be displayed in red, green or blue. This will change the colour of the RGB image accordingly (Figure 1.38). Contrast enhancement and other functions are available via the *Raster toolbox* data stretching. These functions do not alter the data set itself but just how it is displayed.

Raster layers not covering everything below their layer position are uncommon because you usually get a value for every pixel. However, we can set the transparency of individual values and hence display only certain pixel values in our raster, while showing the values of the underlying data below the transparent pixels.

Let us try this using the *unsup_class* file. If you go to the file properties:

```
right-click on layer name > Properties > Transparency (or
double left-click on the layer name)
```

you can set values to *no data*. We have added class 5 to *Additional no data values*; after clicking on *OK*, we can now see the layer values below our unsup_class layer for areas covered by class 5.

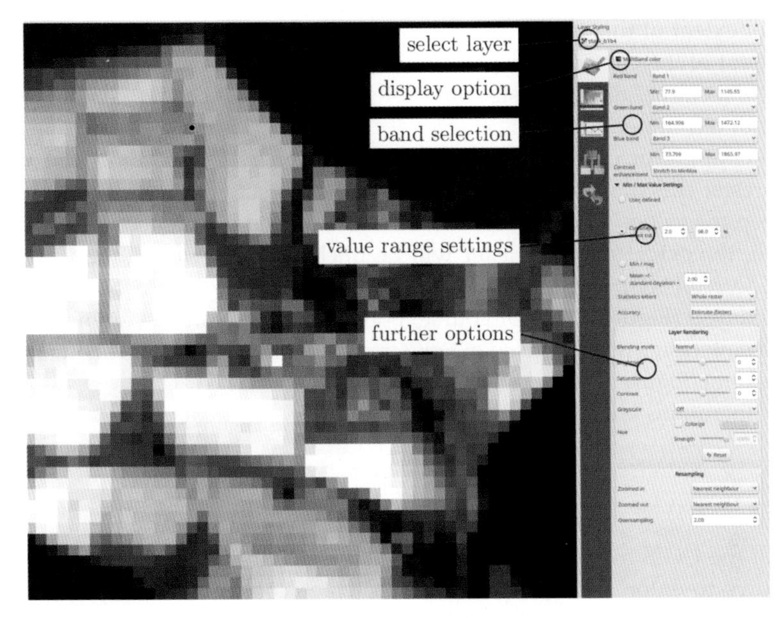

Figure 1.38 Multiband raster layers can be displayed as an RGB image by selecting which band should be displayed in red, green or blue. Single bands of a multiband layer can also be displayed individually by changing the display options to Singleband gray or Pseudocolor.

Raster-vector intersection

A common task applied to raster and vector data is using a vector shape to cut out a section of raster. This kind of operation is called *crop* or *clip*. A similar but slightly different operation is called *mask*. The naming convention of these functions may differ between software packages. Thus, you need to know what the actual function does regardless of its name.

Raster data usually cover much more area than you need for analysis. By cutting them to the actual size of your study area, you reduce file size and speed up any subsequent analysis. Cutting the data (or *clipping* or *cropping*) means cutting out the raster values for a specific extent selected by you (Figure 1.39). Usually, you select a vector file that covers the extent of your area of interest. Thus, you will produce a raster with a reduced extent.

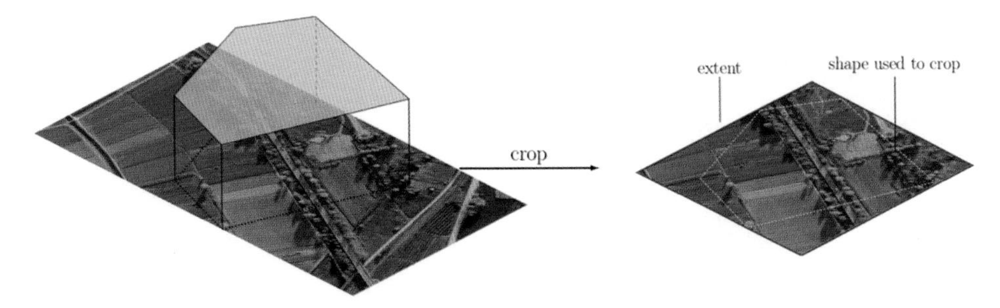

Figure 1.39 The raster data set is cut out to the extent of the vector defining the outline. Cropping or clipping the data set is cutting the raster to the extent of the vector object, not its shape.

The overall extent of the vector file is used to cut the raster. In QGIS, you can instead use the extent of the currently displayed area. However, this makes it more difficult to reproduce the cut on another raster data set since you may change the displayed area; thus, it is not recommended.

While clipping or cropping a raster cuts a raster by the *extent* of the vector file, *masking* a raster cuts the raster by the actual *shape* of a vector polygon. Thus, if you wish to only keep the values of your raster inside a vector shape, you need to apply *masking*. As a result, the extent of your raster will be unchanged; however, the values outside the masked vector shape will be removed and set to 'nothing' (NA). Such values are no longer displayed (Figure 1.40). Of course, both operations can be combined; thus, you can have a new raster with reduced extent (*crop/clip*) and remove values outside a vector shape (*masking*) (Figure 1.41).

These operations are done in QGIS with:

```
Raster > Extraction > Clip Raster by Extent ...
Raster > Extraction > Clip Raster by Mask Layer ...
```

The part of the QGIS interface used to reduce the extent of a vector input to a defined extent allows to enter another spatial object, enter coordinates manually or use the current extent of the map window. Choosing the extent of the polygon vector object results in a rectangular new raster object with the same bands but clipped to the extent of

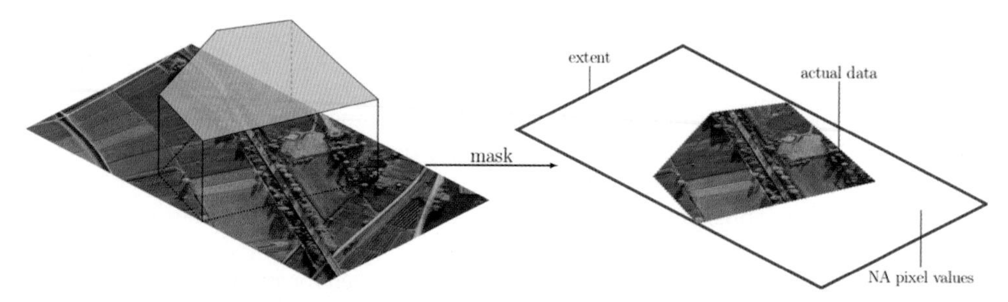

Figure 1.40 All raster values can be set to 'nothing' (NA) by applying a mask based on a vector object. The overall extent of the raster would remain the same but all values outside the vector will not be available and hence not displayed anymore.

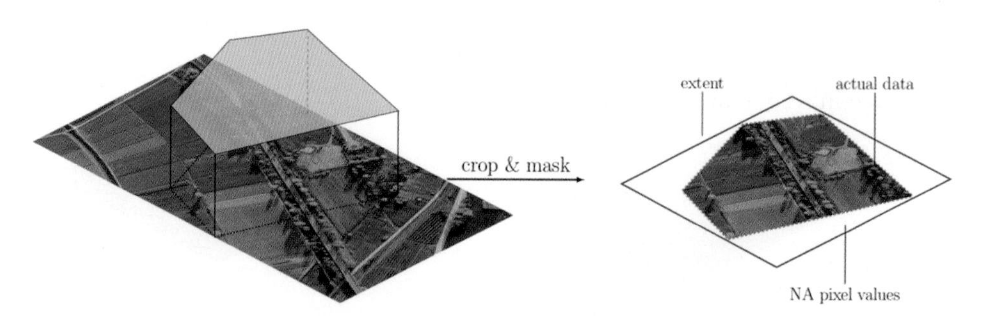

Figure 1.41 Combination of crop and mask reduces the size of the raster to the extent of the vector input and sets all values outside the vector to NA. Thus, only the raster values within the boundaries of the vector file will be available and displayed in the new raster file.

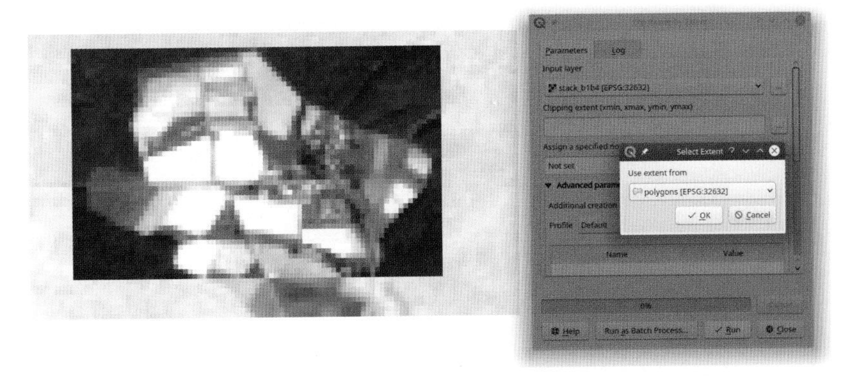

Figure 1.42 The raster layer can be cropped by the extent of the current map view or any spatial object, such as the extent of the polygon. The maximum extent of all objects in the spatial polygon is used to define the cropping extent of the raster. For comparison, the extent of the original raster is shown in the background in shades of blue.

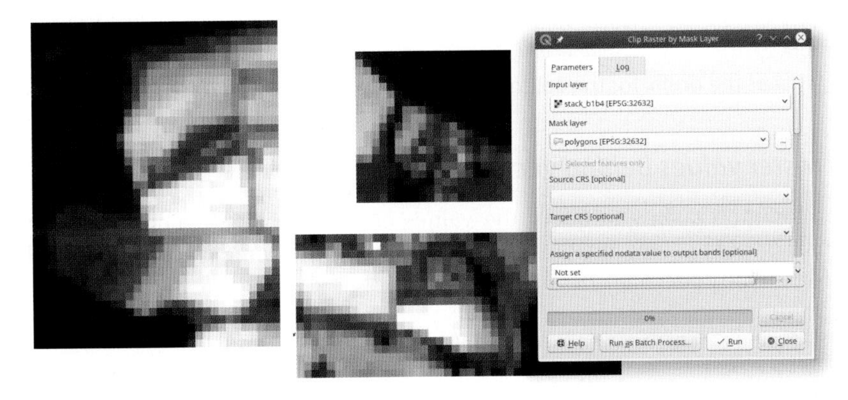

Figure 1.43 Masking all values outside the mask layer of a raster set to nothing while keeping the extent of the original raster.

the polygon object (Figure 1.42). By executing the *Mask layer* operations, you can select a spatial vector object for which all values outside the area of the object are removed from the chosen raster layer. The *Clip raster by mask layer* interface is shown in Figure 1.43; the resulting raster object, where all values in the raster not covered by the polygon shapes have been removed, is also shown.

R COMMANDS

Raster data can be cropped and masked in R using these commands from the raster package:

```
crop()
mask()
```

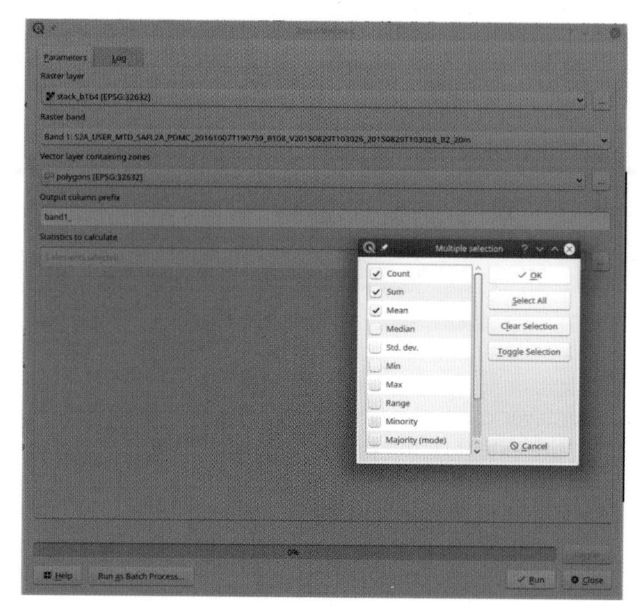

Figure 1.44 Different statistical values of a raster behind a vector can be extracted and added to the table of the vector object. In this example, the values of band 1 are queried, hence band1_ is set as the prefix for the column name inside the attributes table of the vector file.

Query raster values behind vector points

Another common task is to query the values of raster layers behind vector points or polygons. For points, you must install the QGIS plug-in *Point Sample Tool*; for polygons, you must use the *Zonal Statistics* operations:

```
Plugin > Analyses > Point Sampling Tool
Processing > Toolbox > Zonal Statistics
```

Using the raster stack and the spatial vector polygon data sets allows you to query the values of the raster behind the polygon areas. The statistics of the raster values, such as mean or variance, can be selected and will be saved in the attributes table of the vector file (Figure 1.44). It is important to name the *prefix* of the column name inside the attributes table of the vector file appropriately should various rasters be queried.

1.2.2 Projections

One of the key topics in all spatial analyses is the projection used; this topic could cover several books. We do not want to get into the detail but give you (1) an idea of the implications, (2) suggestions about what to use and (3) operations you need to know. Projections are a way to convert the globe's curved surface into a two-dimensional plane. However, to do so we must introduce distortion; the chosen projection will also determine how our globe looks on a plane (Figure 1.45). The impact may be lower at a local scale, but you need to be aware of it.

For any project, regardless of whether you work at a global or local scale, a projection must be decided on and all data must be created or converted (resampled) to the same

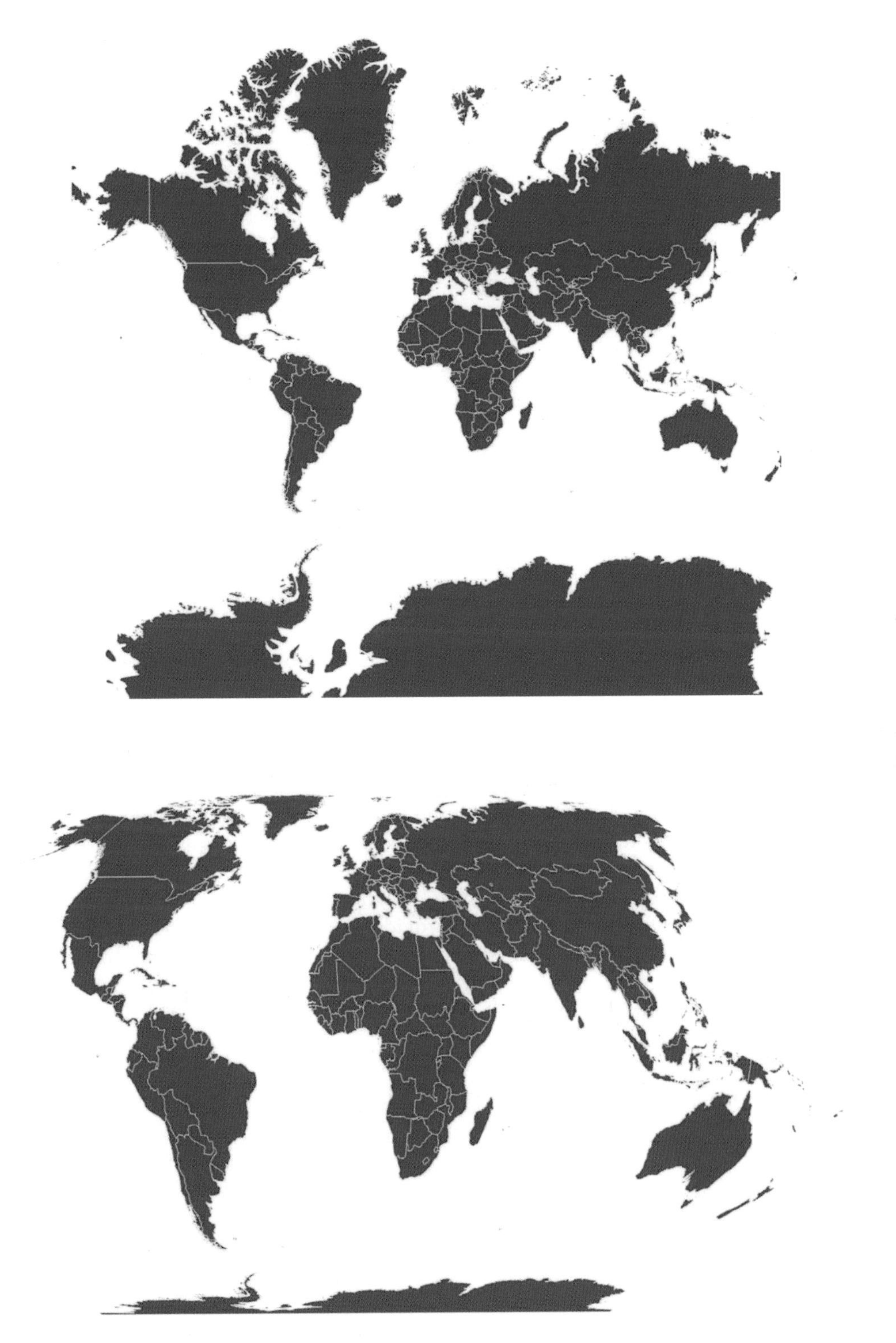

Figure 1.45 Implications of two chosen projections (Mercator and Eckert) on the size, distance and shape of continents and countries.

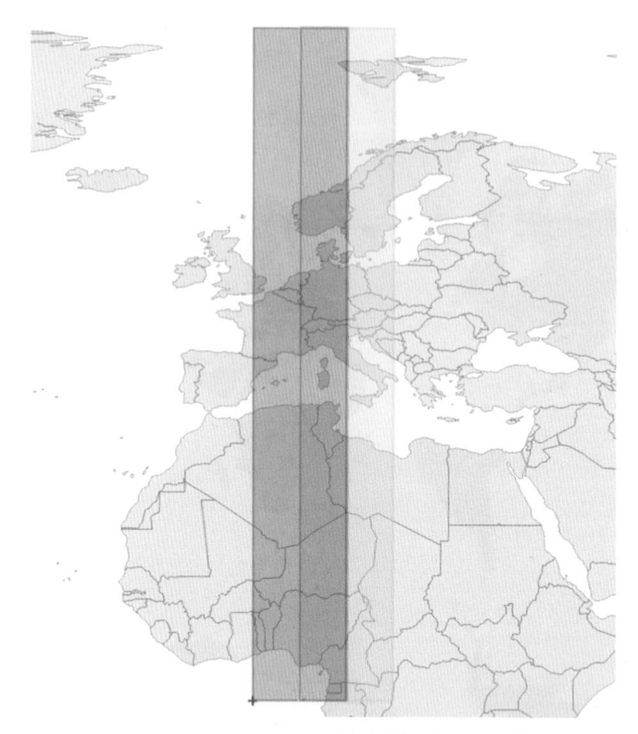

Figure 1.46 QGIS visualizes the spatial coverage of different projections. The coverage of UTM WGS84 for zones 31, 32 and 33N are combined. The projection that covers your local study area should be chosen for all your data.

projection. Moreover, certain projections are created for local studies, others for global ones. Thus, you cannot choose the projection randomly. A recommendation is to use UTM WGS84 for local studies (size of 200 km) and an Albers equal-area conic projection or geographic coordinates (latitude, longitude) for continental or global studies. The Projection Wizard Web page (http://www.projectionwizard.org) recommends suitable projections for your study area. QGIS also provides some visual indications of whether your projection matches the location and extent of your study area. The extent and location of different UTM zones are shown side by side in Figure 1.46. The spatial properties of a certain projection, including its specific parameters, are displayed by QGIS; hence, you can check if it fits with your study area properties.

First, all the data you worked with so far must have the same projection: UTM WGS 84 zone 32. We decided to use this projection for our examples, although we could have used others. Also, all data were created in or changed to this projection for the purpose of the analysis.

To give you a first idea of what the impact of the project might be, we created a grid across our study area with 200 × 200 m cell sizes (Figure 1.47):

```
Vector > Research Tools > Create Grid ...
```

Now we have a regular grid across our study area and the cells are actual squares. If we now change the projection of our view (although not physically reprojecting the data!) by selecting the bottom right button currently, which tells us that we are working on 'EPSG:

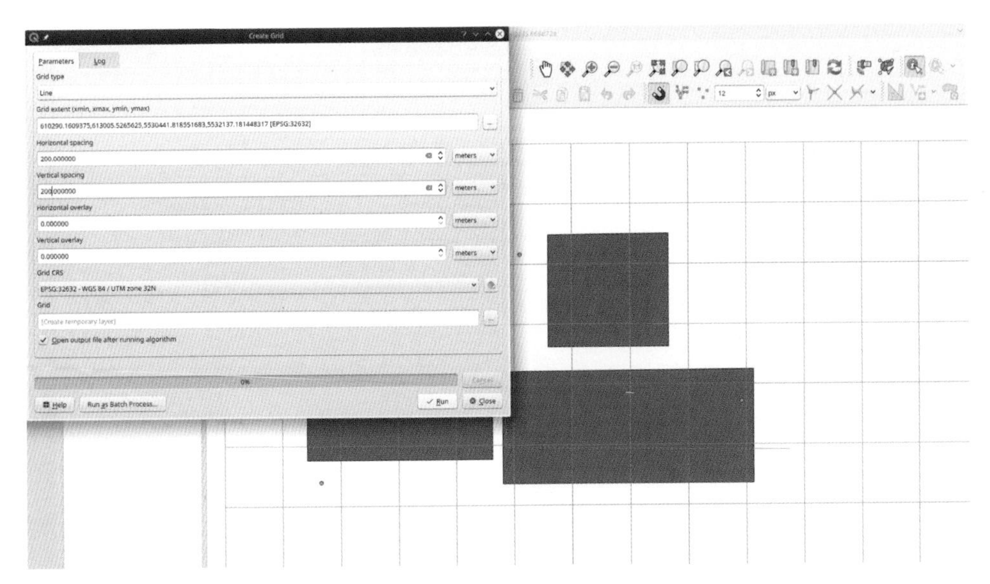

Figure 1.47 The study area of the example vector objects overlaid by a spatial grid of 200 × 200 m.

32632', we can change the projection of our map view to any projection. If we select *Sinusoidal* (IAU2000:39914) or *Albers Equal Area* (EPSG:3574), we have a very good idea of the impact of projections (Figure 1.48). The distorted map, indicated by the orientation and ratio of our grid, is quite different from the original. However, it is still the same data from the same location on Earth, but just in another projection.

Different projections exist for different study area extents, but all come with different (dis)advantages because no projection can represent the globe in a single plane without any distortion. Generally, you will have to settle for one projection. Especially when collecting field data (see Part 2), a projection for the GPS points must be set (and

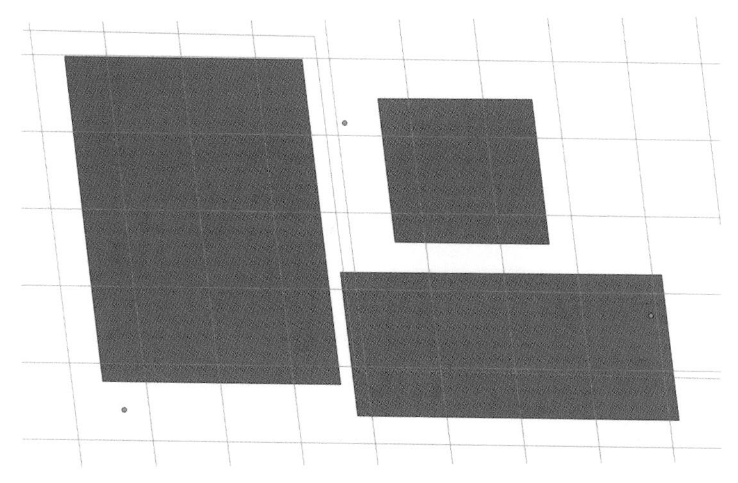

Figure 1.48 The same study area but with a different projection (Sinusoidal IAU2000:39914). The temporary reprojection introduces certain distortion to the grid and vector objects.

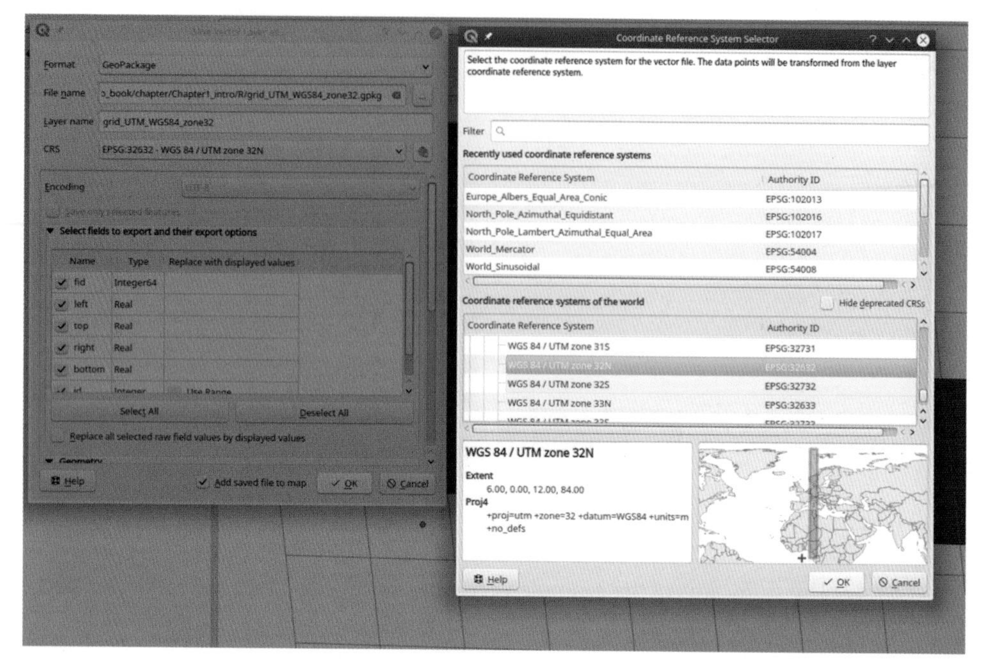

Figure 1.49 Reprojection of a spatial object can be done within the **Save Feature As** operation. Previously used projections, as well as any other projection, can be chosen. The exported object will be available in the defined projection.

recorded). Of course, all data can be reprojected to another projection; however, it is best to minimize such kind of operations if you can. In QGIS, it is quite simple to save an existing object with a new projection by exporting it and defining the new projection (CRS) in the interface (Figure 1.49):

```
Right-click on layer name > Export > Save Feature As ...
```

The same information must be defined when creating a new spatial object. The CRS is by default 'longlat' (geographical coordinates). Here you want to change it to UTM WGS84 zone 32. However, QGIS allows to load data sets with different projections and will reproject them on the fly. This means that the uploaded data will overlap; it will look like they have the same projection while physically they will not. This feature of QGIS is very useful but it can also lead to problems. Therefore, we recommend that you check the projections and reproject if necessary. R, for example, requires a 100% match of the projection string for any spatial analysis.

R COMMANDS

Reprojection of raster and vector data is done within R using the 'rgdal' package:

```
projectRaster()
spTransform()
```

QGIS INFO

A fun tip for QGIS (also called Easter egg) is to enter 'bored', 'retro' or 'dizzy' in the coordinate field at the bottom of the QGIS window. Just replace coordinates such as '612496,5530472' with one of these words, but save your QGIS project beforehand.

1.3 Next steps

This chapter has provided you with the basic principles of spatial data and its practical application. In the next chapters, actual data are used and a variety of operations are performed on them, such as reprojecting or cropping. Naturally, you will be faced with different challenges when working with actual data for a specific project but applying the skills you have learnt and exploring the spatial operations will allow you to solve them. Of course, there are many more operations, information and processes to learn before spatial data handling is mastered to its full extent. However, this and the following chapters provide a sound foundation that allows you to start exploring spatial research on your own.

Part I

Data acquisition, data preparation and map creation

In this part of the book, all topics that need and should be covered before any field trip and data analysis are addressed using actual data from a real landscape example.

We start by outlining potential data sources for informative overview data and real-world analysis and continue by applying the first necessary data adjustments, such as cutting the data to the study area. Furthermore, spatial data modifications that might be necessary for the analysis are addressed.

We conclude Part I by creating maps of the acquired data sets. Maps are an important part of any stage of a research project: they are necessary for presenting your results at the end of the project, but also before any fieldwork to get a good overview of the study area as well as guides for the actual field trip.

2. Data acquisition

In Chapter 1, we outlined the fundamental properties of spatial data and spatial operations using artificial data sets. An actual research project requires real environmental information. However, using real data involves a range of different challenges starting with the data availability.

In general, before you start a project you need to know:

- Which area needs to be covered (extent).
- When (which year, season or day) environmental information is needed.
- What kind of data are needed to answer the research question.
- For what purpose you need which data set (mapping or analysis).

All these questions must be considered beforehand because you want to assemble a range of environmental data for a specific task within your *area of interest* (AOI) and generally because environmental information is needed for a specific time (season or year). If you wanted to compare environmental information at the time of your fieldwork, then you need to temporally match remote sensing data acquisitions. However, if your aim is to analyse land cover conversion, you would look for data sets of the same time of the year but in different decades to map, for example, deforestation. Linked to this issue, which type of data you need to answer your research question must be addressed. For deforestation on a small scale, you might want to investigate satellite data such as Sentinel-2 or Landsat data, whereas Moderate Resolution Imaging Spectroradiometer and Sentinel-3 data are the best choice for mapping changes on a larger scale. Additional relevant data, such as road network, settlements or national park boundaries are usually provided as vector data. All these different data sets are usually combined in your analysis; therefore, they must match spatially and temporarily.

Due to these considerations, this chapter covers the data acquisition of different types of spatial data for your AOI. In general, data differ in format (vector or raster) and information content.

2.1 Spatial data for a research question

Spatial data are recorded, processed and published by a multitude of data providers. These range from open-source development communities (e.g. OpenStreetMap (OSM)) over publicly funded data services (e.g. European Space Agency (ESA) Copernicus or NASA's Earth Observing System) to private companies (e.g. Planet, Mapbox, Google,

Amazon). While the data products offered by such providers all share the spatial component, they differ, for example, in:

- Type (vector or raster data).
- Class (e.g. thematic or imagery data).
- Quality (e.g. clouded).
- Resolution (the level of detail, e.g. spatially, temporally or spectrally).
- Geographical extent (e.g. local, continental, global).
- Temporal coverage (e.g. static data representing a specific instant in time; daily, weekly, monthly or otherwise multitemporal data).
- Level of processing (e.g. raw sensor data, data representing physical measures, informative maps).
- Purpose (e.g. scientific temperature model data or daily use traffic maps).

The general relevance of a spatial data product for a certain research question can be evaluated based on characteristics that can be accessed and evaluated before acquiring the product. For example, spatial data products with a low level of specialization, such as satellite imagery without spectral bands that are just RGB images (such as the images used by popular online mapping services), potentially contain useful information. However, they require higher data processing efforts to turn the data into quantitative results that can be used to investigate the research question. Features in the landscape need to be manually extracted (digitized). On the other hand, spatial data products that represent a highly specific quantity, for example, air temperature or soil moisture, require less processing but are only suitable for a narrow range of research questions. Similar 'trade-offs' may emerge for other characteristics, for example, data storage requirements depending on type, geographical extent and temporal coverage.

Moreover, the required spatial information might not be available or might have limitations. Cloud cover or inadequate spatial or temporal resolution, data quality or geographical extent might render the data unusable for your research project. In such cases, it may be useful to acquire and process lower resolution data, for example, spatial resolution to derive the required measure. Moreover, data sets such as OpenStreetMap may contain valuable information needed to investigate a research question such as buildings but are mixed with other features that are not needed, for example, roads. In this case, some geoprocessing operations are needed to isolate the required feature.

INFO

For an overview of different spatial data products available on the Web, visit http://remote-sensing-biodiversity.org/resources/, whose product and Web portal link list is regularly updated.

Remote sensing imagery is derived through either *passive* or *active* sensors. Multispectral data are provided by passive sensor systems, whereas radar data belong to the latter (Figure 2.1). While passive systems rely on light emission and are hampered by clouds or any atmospheric disturbance, active systems are independent of both issues. In this book, only data from passive sensors are used.

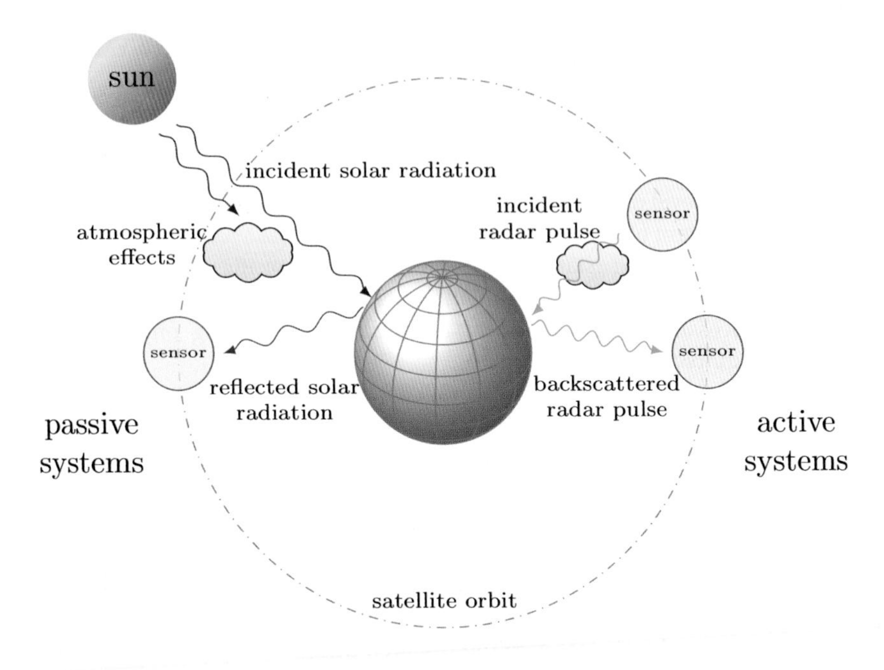

Figure 2.1 Different data types have different limitations. Atmospheric disturbances, such as clouds, hamper optical data acquisition, whereas active systems are unaffected by cloud cover in general. Active sensors, such as radar, come with their own challenges. In this book, only data acquired using passive systems (multispectral imagery) are used.

2.2 AOI

Before searching, downloading and displaying spatial data products, it is useful to define the AOI. AOI is used to limit all operations to a geographical extent. Often, it can be advantageous to select an AOI that covers the study area and add a certain margin. It may be better to acquire a slightly larger extent instead of having problems at a later stage with study sites being too close to the edge of the AOI. The analysis conducted in this book is based on a predefined AOI stored as a vector file in the accompanying data folder. Open a new QGIS session and load the AOI into QGIS by selecting:

```
Layer > Add Layer > Add Vector Layer ...
```

The QGIS *Data Source Manager* for vector data opens (Figure 2.2). Select the **...** icon to the right of the input field named 'Vector Dataset(s)' to open the file manager of your operating system and navigate to the directory of the folder accompanying this book. Navigate to the folder '.../data/Chapter_2/AOI/' and select the file 'aoi.shp'. The .shp file format comes with a variety of additional files; are all needed, otherwise the vector information will no longer be useable. GeoPackage (.gpkg) is an alternative format that comes as a single file. In Chapter 3, we cover different file formats in more detail.

Select *Open* to return to the *Data Source Manager*. Select *Add* to add the AOI to your QGIS session. Select *Close* to close the *Data Source Manager*. Alternatively, you can drag and drop the file to the *Layer* section of QGIS to add it.

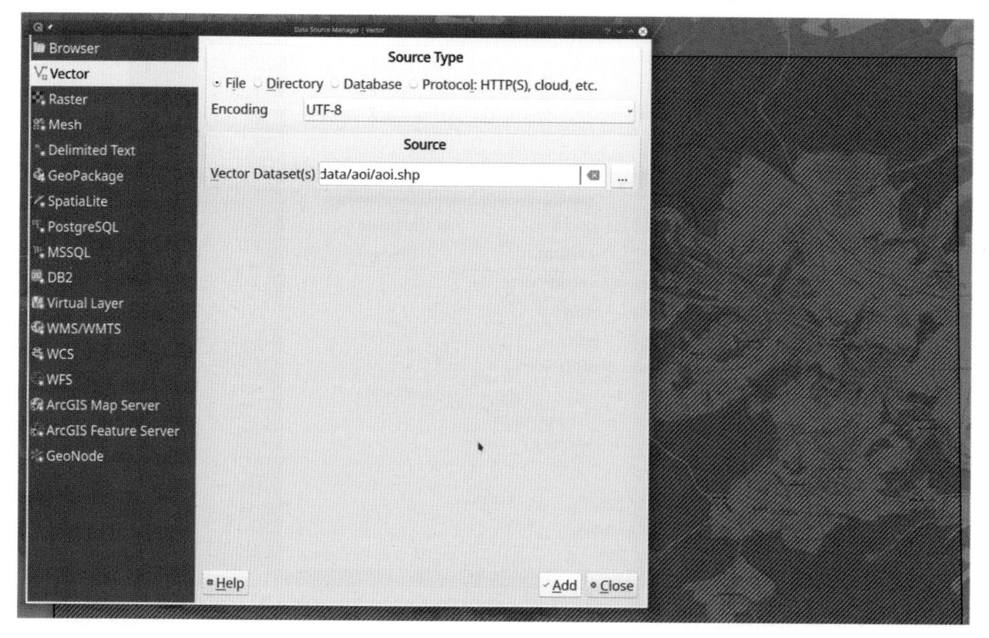

Figure 2.2 The vector data register of the QGIS Data Source Manager, which can be used to load vector data into QGIS. The imported polygon is shown on the map with the OpenStreetMap data in the background.

QGIS INFO

Creating your own AOI can be done easily through Layer > Add Layer. Creating new vector polygon layers is covered in detail in Part 2.

Before we continue, have a look at the projection of the newly added AOI layer. For this, select the layer and open the context menu by right-clicking:

```
Properties ... > Information
```

and check the coordinate reference system (*CRS*) field under *Information from provider*:

```
EPSG:3857 - WGS 84 / Pseudo-Mercator - Projected
```

As you can see, the AOI is stored as a variant of the *Mercator* projection called *Pseudo-Mercator*. This projection is also often referred to as the *Web Mercator* projection since it is the standard projection used by nearly all map Web services, such as Google Maps or OpenStreetMap. In addition, the *CRS* field shows that the *World Geodetic System 84* (*WGS 84*) ellipsoid is used as the geodetic reference system.

It is good practice to always review the projection of a newly added layer. Since the AOI layer has been the first layer added to this QGIS session, QGIS will by default use its projection as the general projection of the running session. This means that further layers added to the session will be reprojected on the fly to the projection we have just reviewed, in case they are physically stored in a different projection. In our case, this is

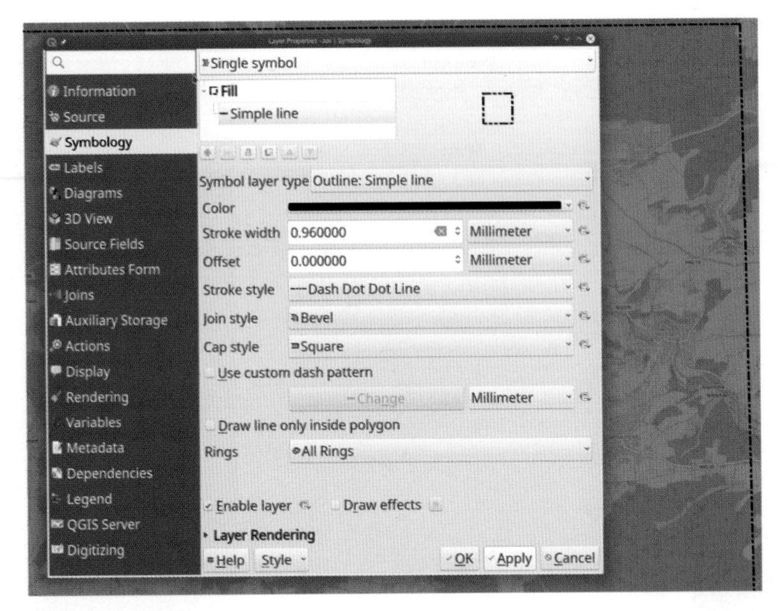

Figure 2.3 The Symbology tab with the vector Layer Properties window. The background shows the resulting vector style.

useful because the map data products we want to add to the QGIS session in the following sections come with the *Web Mercator* projection. Additional data that are added may have a different projection; nevertheless, they will be reprojected on the fly. How to deal with downloaded data of diverging projections is addressed in Chapter 3.

Before closing the *Layer Properties* window, navigate to the register 'Symbology' (Figure 2.3). Here, the colouring of the AOI can be edited. By default, QGIS displays polygon vectors (like the AOI) with a *Simple fill* symbology. Thus, the AOI polygon appears in a randomly selected colour that is not transparent and will cover all subjacent layers that might be added. Since we are interested in visualizing the boundaries of the AOI and not its fill, we adapt the symbology by selecting the 'Simple Fill' field. Then, we open the *Symbol layer type* drop-down menu and select 'Outline: Simple Line'.

TASK

Change the vector styling to different fill and line types. Change the colours and translucency. Explore how to best display such data.

Choose a colour visible on your background, for example, black or orange, by clicking on the down-facing arrow to the right of the *Color* field and choosing the colour from the *Standard colors* section. Lastly, you might want to increase the stroke width by changing the number in the *Stroke width* field or the *Stroke style*. Finally, click on 'OK' to save the changes and return to the QGIS main window. The QGIS viewer now displays the AOI boundaries that will be used to access the spatial data products. Having a rectangle on a white background is not too meaningful, hence relevant background information can be added.

QGIS INFO

Keep all layers that will be added in the course of this session in your current QGIS session. You can deselect layers that should not be visible for the current task. In Chapter 3, you will learn how to transform all layers added in this chapter into equal projections, how to crop them to the AOI and how to store them.

2.3 Thematic raster map acquisition

To get an overview of the study area, it is useful to acquire one or several base maps to be displayed in the background. Depending on the map type and/or the service provider, base maps are offered both as raster and vector data. Raster base maps are very suitable for creating custom maps (e.g. of a study area) but are more difficult to use for analysis because individual map features cannot easily be extracted and are merged as a raster. A widely used, open-source map service is OpenStreetMap, which offers global thematic maps on land use, traffic networks, administrative boundaries, landmarks and many more.

In QGIS, map data can be downloaded and displayed using the *QuickMapServices* plug-in, which can be installed by searching for *QuickMapServices* and select *Install Plugin*:

```
Plugins > Manage and Install Plugins ...
```

The actual data can then by loaded by navigating to:

```
Web > QuickMapServices > OSM
```

and selecting *OSM standard*. A global thematic standard map layer is added to the *Layer* section in QGIS (Figure 2.4). If the map layer is displayed above the AOI layer, just drag and drop the AOI layer to the top position in the *Layers Panel*.

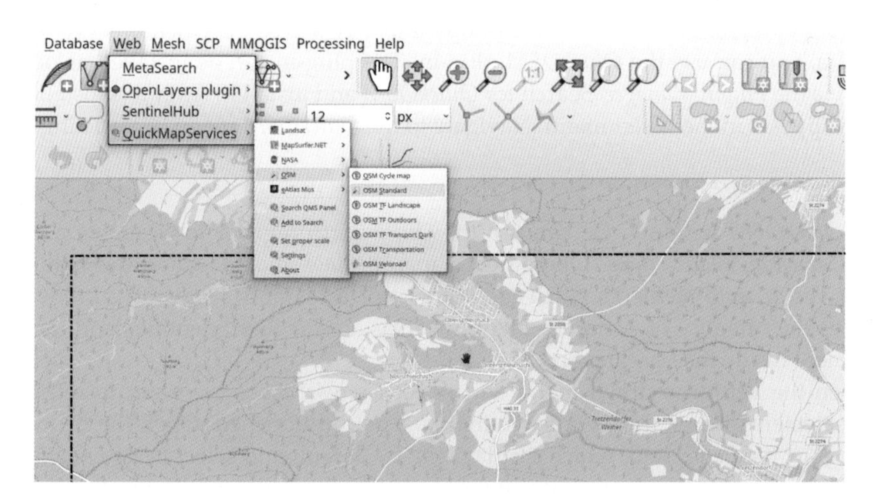

Figure 2.4 Adding an OpenStreetMap standard base map using the QuickMapServices plug-in. The resulting map is displayed in the background with the AOI outline as the line symbol.

TASK

Explore the map layer we just loaded. Use the mouse to click, hold and drag the map for navigation or navigate using the arrow keys. Use the mouse scroll wheel to zoom in and out. Check out the region covered by the AOI.

The newly added layer is a *raster* layer and you can zoom into the map and navigate from a global to a highly detailed local extent. However, when you do this, you will notice that each time you zoom in or out, the layer's appearance changes slightly. For example, map labels appear, disappear or relocate, colours are adjusted and map details vanish or emerge. Unlike when you zoom into a classic image file (e.g. a photo) with standard image viewing software on your computer, the OpenStreetMap layer you are looking at does not represent a single raster image covering the whole world at the highest possible resolution. Instead, it is composed of multiple images that only cover local extent with different degrees of detail depending on the zoom level you choose. These zoom- and extent-related raster images of base map layers are referred to as *map tiles* or *slippy map tiles*, since you can move and scroll through the map without noticing the edges of each tile.

However, providing this spatial information as raster data prevents you from being able to select single spatial features, for example, all roads or a single building. This can be achieved if the same data are downloaded in their original format, that is, as vector data.

2.4 Thematic vector map acquisition

Instead of accessing an OpenStreetMap map rendered as raster map tiles, the underlying map features used by the OpenStreetMap server to compose these map tiles can be retrieved independently and individually as vector data. Such data can then also be used to create your own maps with individually chosen spatial features or add certain features on top of raster base maps, such as points, lines or polygons. Most importantly, this vector information can be used for analysis, for example, to measure distances between points, buffer regions or any other vector operation. To access the OpenStreetMap vector map data, you can use the *QuickOSM* plug-in. The plug-in can be installed as outlined earlier via the *Plugin Manager*. Search for *QuickOSM* and install it. After it is installed, the plug-in appears in the *Vector* menu. Navigate to the *QuickOSM* menu by selecting:

```
Vector > QuickOSM > QuickOSM
```

In the *QuickOSM* window, we can use the *Quick query* register to assemble a custom OpenStreetMap vector map. OpenStreetMap uses *tags* that describe the features of specific vector map elements. Such tags consist of a *key* and a *value* item. A *key* item describes the category of a map element, whereas a *value* item identifies and describes the element inside the *key* category.

Open the *Key* drop-down menu and select *landuse*. Since we want to retrieve all *landuse* elements inside the AOI, leave the *value* field untouched. Open the drop-down menu below the *Value* field, which displays '*In*' by default and select *Layer Extent*. Select the AOI layer from the drop-down menu to the right. This results in all data being downloaded for the extent of that layer. Executing *Run query* retrieves the data from the OpenStreetMap server. After the query has been executed, three land use layers will

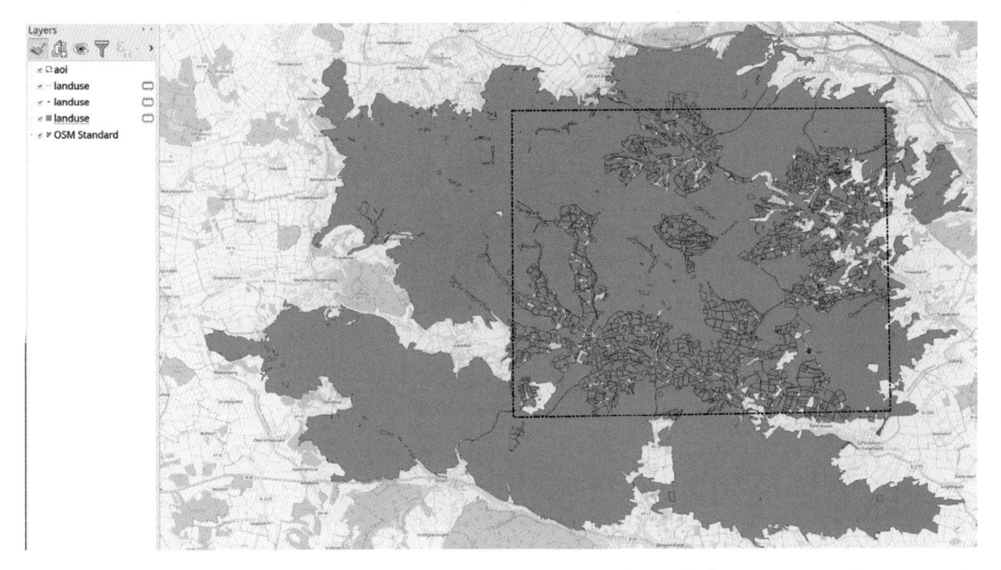

Figure 2.5 The QuickOSM query results are shown in purple, with the OSM raster base map in the background and the outline of the AOI as a vector line.

appear in the *Layer* section (polygon, line and point). The *QuickOSM* window can now be closed and the newly added layers can be viewed and inspected.

Move the AOI layer to the top position in the *Layer* section. You may recognize that some large land use polygons extend the AOI (Figure 2.5). OpenStreetMap returns all map elements that cover the AOI, but their extents are also returned. First, focus on the polygon layer. Deactivate the line and point land use layers by clicking on the tick located to their left the *Layer* section.

To get a better understanding of the loaded OpenStreetMap data, it is useful to modify its colouring. This can be achieved by double-clicking on the polygon land use layer to open the *Layer Properties* and navigate to the symbology register or by activating the *Layer Styling* panel.

Change the styling from *Single Symbol* to *Categorized* and select *landuse* in the *column* field. Select *Classify* at the bottom of the register to automatically define all land use classes contained in the column *landuse*. Although the colours are selected randomly and may not necessarily fit the classes, the new colours can be used to separate different classes. Changing some of the classes to more intuitive colours may be helpful to visually understand the meaning of the classes shown (Figure 2.6).

TASK

Assign colours to each class that match the class meaning (e.g. assign dark green for forest). Explore other symbol layer types beside 'Simple Fill' such as 'Shapeburst Fill' or fill the polygons with colour gradients or diagonal lines.

Compared to the previously accessed OpenStreetMap raster map tiles, the newly added OpenStreetMap land use vector layer includes less information. This is because

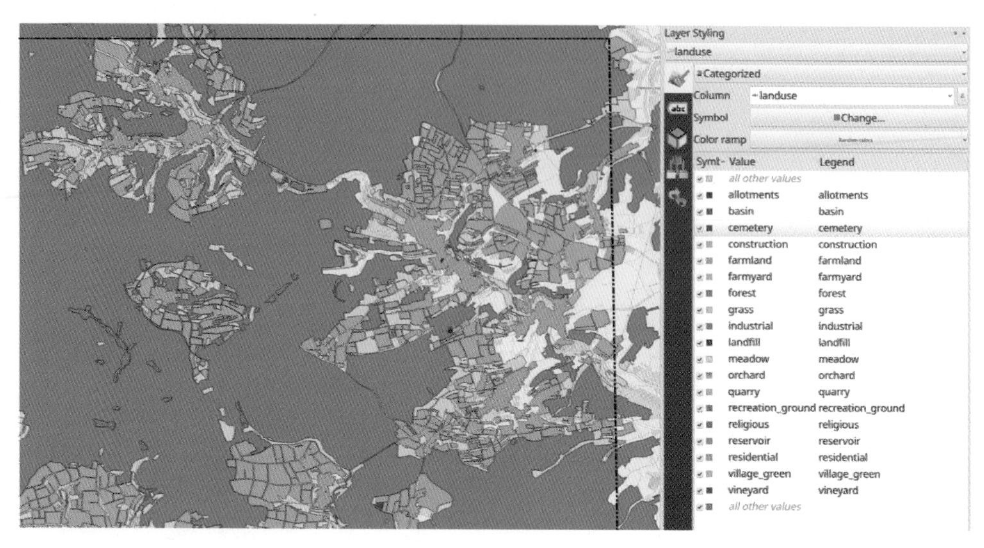

Figure 2.6 OpenStreetMap land use polygons shown in different colours depending on the land use category. The AOI extent is shown as a line.

we have accessed only certain categories (*tags* of a single *key*). However, standard OpenStreetMap raster tiles are rendered from a multitude of keys, including all kinds and types of data. The styling also differs. While we have used simple colouring to differentiate land use classes, the OpenStreetMap server uses specifically designed textures and symbols for each *tag*.

QGIS INFO

If you want to learn more about which tags are used by OpenStreetMap, visit the Standard tile layer wiki (https://wiki.openstreetmap.org/wiki/Standard_tile_layer). Under the Map key subsection, you can find links to lists of OpenStreetMap tags consisting of a key and a value, along with the associated symbols and textures used to generate the raster tiles. You can use those keys and values with the *QuickOSM* plug-in to import the associated elements into QGIS as vector data.

Theoretically, you could use *QuickOSM* to add all map elements included in the OpenStreetMap raster map tiles to mimic the latter as a custom vector map. Generally, it might be more useful (and saves space) to retrieve only those map elements you need for your specific map.

TASK

Explore the potential of *QuickOSM* by adding more elements to your QGIS session. Repeat the procedure outlined for the keys 'admin_level', 'place' and 'highway'. These, in addition to the already accessed 'landuse' key, are needed for the analysis examined in the following chapters.

Table 2.1 Layers required for the analysis used in the book

Key	Layer type
landuse	Polygon
admin_level	Polygon
place	Point
highway	Line

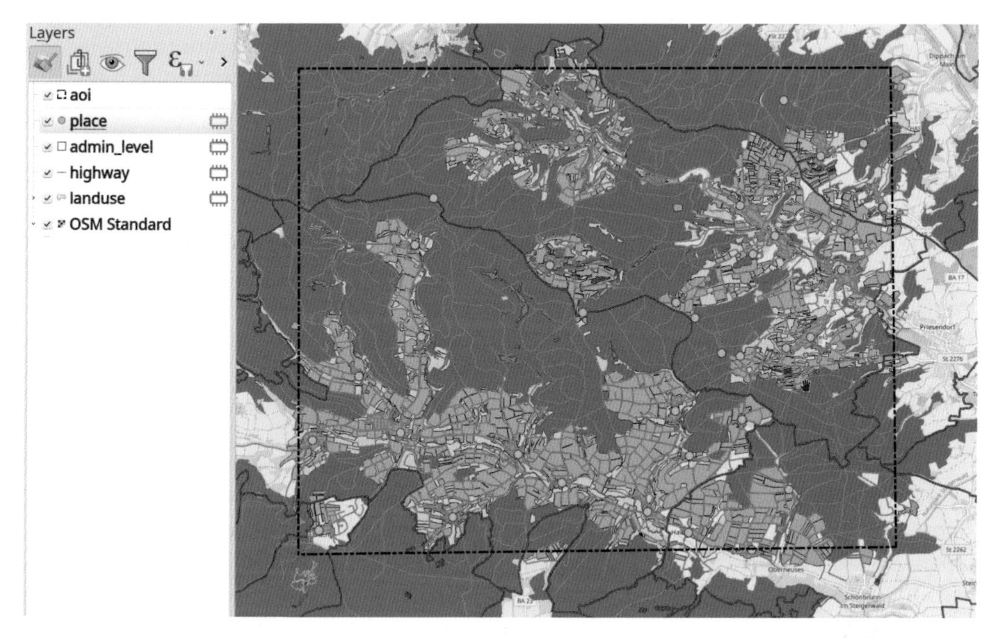

Figure 2.7 The QGIS session including all layers of interest to be used in the analysis. Redundant vector types are removed, and the styling of each vector object is adapted.

For each key, *QuickOSM* will load a point, line and polygon layer, which are unimportant if actual information is provided. For the analysis in this book, only some of these layers are needed (see Table 2.1):

Therefore, you may want to remove the empty or unnecessary layers for each *key* before proceeding and keep only the layers of interest (Figure 2.7).

2.5 Satellite sensor data acquisition

The thematic maps provided by OpenStreetMap combine data from a multitude of sources, including various databases on geographical names and attributes, authoritative map publications, satellite and aerial imagery and more. These data products have been curated, processed and visualized with the goal of creating a thematic, human-interpretable, informative map. Through such valuable addition, original sources can be merged into a refined, unique product that represents a higher level of processing.

While such a product covers specific informative content, filtered for a certain purpose, spatial data products involving lower levels of processing, such as sensor imagery, contain an unfiltered, unspecified pool of information. For example, a satellite image with four bands representing the blue, green, red and near-infrared range of the electromagnetic spectrum can be used to extract certain landscape features, such as buildings, roads and railways. However, the same image can also be used to detect vegetation, estimate photosynthetic activity or classify land use. Thus, spatial data products involving a low level of processing allow the individualized, study-specific extraction of various information that may not be covered by higher-level, more specific products such as thematic maps.

Satellite sensors have different technical characteristics (e.g. spectral or spatial resolution as outlined in Chapter 1). These characteristics may have advantages and disadvantages when investigating a certain research question. Thus, a thorough selection of a (or multiple) suitable sensor(s) based on the characteristics and context of the study question, area and time span is very important.

2.5.1 Sentinel-2 data

For the analysis conducted in this book, we use imagery from the multispectral imagery (MSI) sensor orbiting the Earth on board of Sentinel-2, which is operated by ESA within the Copernicus Programme. The Sentinel-2 MSI imagery covers a wide spectral range with 12 bands overall; bands 2, 3, 4 and 8 (red, green, blue and near-infrared) are acquired with a spatial resolution of 10 m (Figure 2.8). These bands are of interest for our analysis and a spatial resolution of 10 m is appropriate for our purpose of mapping the land cover information for this area.

Sentinel-2 data are provided freely by ESA through a Web service that can be used by third-party software or accessed through the Copernicus Open Access Hub (https://scihub.copernicus.eu/dhus/#/home). For our data acquisition, we are going to use the Open Access Hub to create a Copernicus account. If you already have a Copernicus account, you may skip this step. Otherwise, go to *Login > Sign up* (represented by the human silhouette flanked by a?) on the Web page and follow the instructions to sign up and confirm the registration. After you have successfully registered, you can close the hub, since it will not be used in this exercise. However, you can search, query and download data directly from this hub if you so wish.

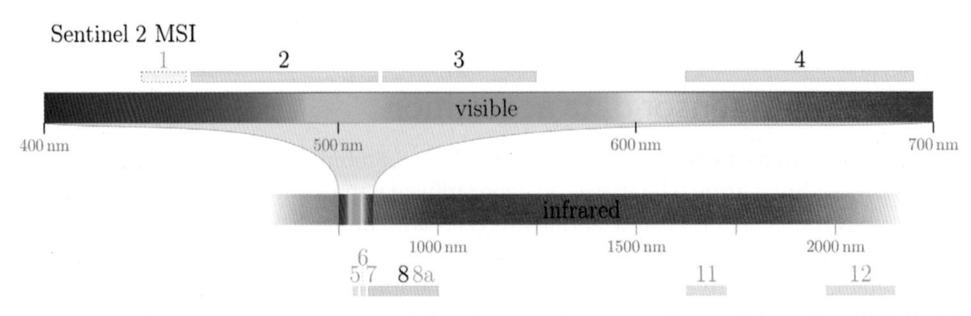

Figure 2.8 The Sentinel-2 bands used for the analysis are in the visible and near-infrared spectrum.

QGIS INFO

A variety of extensions use Web services to query, preview and download spatial data: openEO, Sentinel Hub, QuickMapServices or OSM Downloader.

The *Semi-Automatic Classification Plugin* (*SCP*) can be used to query, preview and download Sentinel-2 MSI imagery. Besides its main features for classifying imagery directly in QGIS, *SCP* provides tools that can be used to acquire imagery from various satellite sensors, including Sentinel-2.

Install this plug-in via the *Plugin Manager* by searching for *Semi-Automatic Classification Plugin*. After you have successfully installed *SCP*, the user interface of QGIS will display following new elements:

- The *SCP* toolbar above the viewing pane.
- The *SCP* dock on the left, below the *Layer* section.
- The *SCP* drop-down menu in the menu bar at the top of the QGIS window, beside the *Processing* drop-down menu.

Open the *SCP* download panel by selecting:

```
SCP > Download products
```

The *SCP* main panel will open, displaying the *Download products* section and the *Search* register by default. First, login via the *Login data* tab and enter your login credentials (username and password) in the *Login Sentinels* section (Figure 2.9).

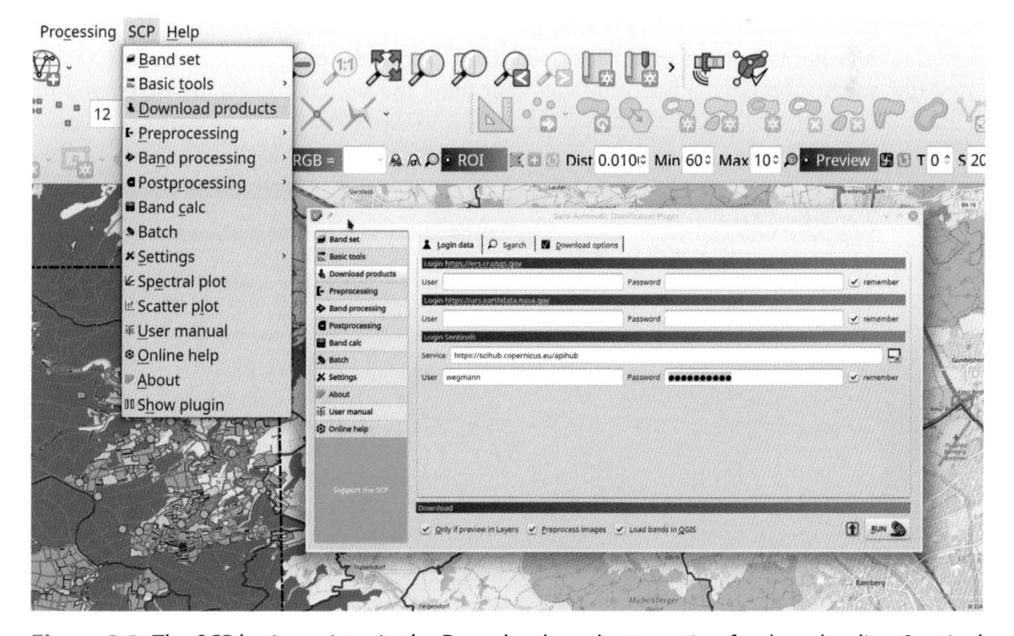

Figure 2.9 The SCP login register in the Download products section for downloading Sentinel data.

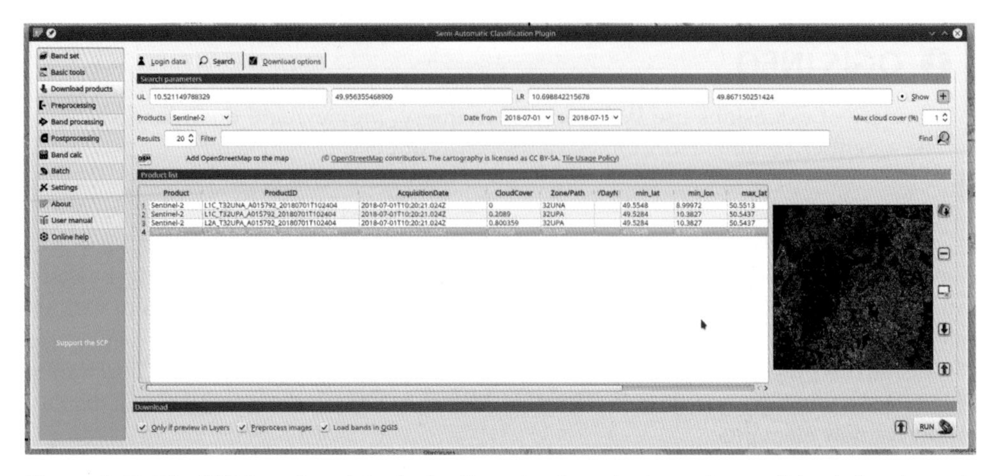

Figure 2.10 The SCP Search register in the Download products section and the defined search parameters including the product list returned by the search and a preview of the selected scene.

Now, we are set up to access the Copernicus services. Switch back to the *Search* interface to enter some search parameters. First, we define a bounding box based on our AOI. Select the *plus* symbol on the right, then switch to the QGIS main window. Mark the upper left corner of the AOI with a left-click and then mark the lower right corner of the AOI with a right-click. A polygon will appear, indicating the selected area. At the same time, the upper left and lower right coordinates are added automatically into the *Search* parameters section. Now, select 'Sentinel-2' as the *product*. Choose '2018–07–01' as the start date and '2018–07–15' as the end date. Lastly, set the field *Max cloud cover* to 1 since we want to find imagery that is not or only partly covered by clouds. Setting the maximum cloud cover to 100 would provide all available images. Click on the *Find* button to the right below the field *Max cloud cover* to start the query (Figure 2.10).

After a few moments, a list of images that meet the search criteria are displayed in

Figure 2.11 The SCP Download options register in the Download products section; the selected bands required are shown.

the *Product list* section. When you select a record, a preview is displayed on the right. The *Product list* table consists of several columns. By clicking on the *minus* button to the right of the preview window, it is possible to remove records from the list. For our analysis, we are only interested in *Level 2A* imagery that is atmospherically corrected. In addition, we only want to retrieve the *T32UPA* tile that covers our study area. The *UNA* tile covers an adjacent area. Thus, select all records except for the record with the *ProductID* starting with *L2A_T32UPA_* and remove them. Before starting the download, make sure you deactivate the *Only if preview in Layers* and *Preprocess images* options (since *Level 2A* images are already preprocessed). Lastly, switch to the register *Download options* and go to the *Sentinel-2 bands* section and activate bands 2, 3, 4 and 8 (blue, green, red and near-infrared), that is, deactivate all other bands (Figure 2.11).

To download the imagery, select *Run* and choose a folder where the imagery should be saved to. The download will start after a few moments and your QGIS session will be unavailable until the download is finished. Close the *SCP* window after the download has finished. Four raster layers have now been added to your layers, each representing one band. They are saved as *JPEG 2000* (.jp2) files in the defined directory.

TASK

Change the layer styling from grey to pseudocolour. Move the layers to the top or (de)activate layers to inspect each of the new layers. Right-click on one of the bands and select *Zoom to layer* to view the full extent. Check the projection of the downloaded data via the *properties* interface.

To locate the region we are interested in, it is useful to display the AOI on top of the Sentinel band layers (Figure 2.12). We have now previewed, selected and downloaded

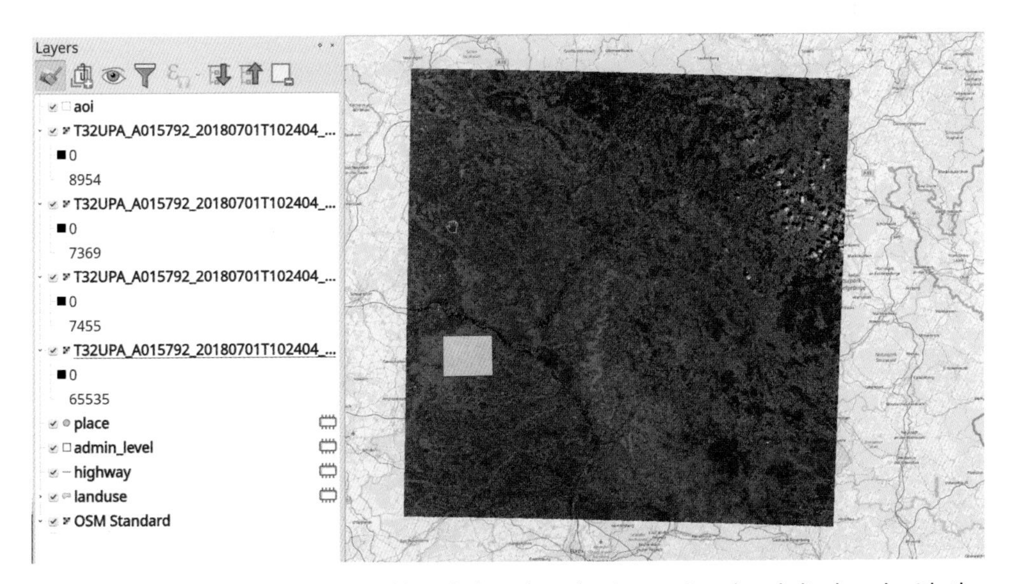

Figure 2.12 Downloaded Sentinel bands listed in the *Layers Panel* and displayed with the study area extent vector on top.

Sentinel-2 MSI multiband data for one date (1 July 2018) using *SCP*. You may notice that the downloaded area covers a notably larger area than we want to investigate. This is one of the reasons why further processing of the downloaded data is needed. We will address this in Chapter 3.

> **TASK**
>
> *SCP* also supports downloading data from the U.S. Geological Survey (USGS) and NASA. Register at the USGS EROS Registration System (https://ers.cr.usgs.gov/login/) and NASA Earthdata (https://urs.earthdata.nasa.gov); then go to the *SCP* login and enter your credentials. Download the Landsat data (L8 OLI/TIRS) with bands 2 (blue), 3 (green), 4 (red) and 5 (near-infrared) for a date close to 1 July 2018. Compare both Sentinel-2 MSI and Landsat 8 OLI bands. Can you see any differences?

2.5.2 Shuttle Radar Topography Mission (SRTM) digital elevation model (DEM) data

In addition to Sentinel-2 data, DEM data are required for the analysis conducted in this book. DEM data are raster data containing single values per pixel that represent the elevation of that pixel. A popular DEM data set that is widely used in remote sensing science is the SRTM DEM. SRTM was a Space Shuttle Endeavour mission (STS-99) conducted in February 2000, which collected synthetic-aperture radar data using a C- and X-band instrument of the entire globe. The C-band data have since been processed by NASA's Jet Propulsion Laboratory to derive global DEMs. The newest version, SRTM DEM version 3, offers a spatial resolution of 30 m.

Figure 2.13 The SRTM Downloader plug-in for QGIS.

In QGIS, SRTM DEM data can be simply downloaded using the *SRTM Downloader* plug-in (Figure 2.13). Install the plug-in using the *Plugin Manager*. Before opening the plug-in, make sure that the map canvas is zoomed to the AOI layer by right-clicking on the AOI layer and selecting *Zoom to layer*. Then, open the *SRTM Downloader*:

```
Plugins > SRTM Downloader > SRTM Downloader
```

Select *Set canvas extent* and choose an output path before selecting *Download*. You will be prompted to enter your USGS EROS Registration System credentials before the download proceeds. Since the AOI of the project covers only a small area, a single SRTM DEM tile will be downloaded and added to the *Layer* section. By default, it will appear in grey. Black indicates valleys, whereas white indicates higher elevation, such as hills or mountains.

2.6 Summary and further reading

The data acquisition outlined in this chapter covers just a small fraction of potential data types and sources. Besides multispectral reflectance and elevation data, many more data sets exist that are worth exploring, depending on the field of research or the analysis goals. These include other sensor data, for example, synthetic-aperture radar or hyperspectral data, as well as products derived from such data by using complex models, for example, air temperature or land cover data. In this chapter, all data sets were downloaded using dedicated QGIS plug-ins. If no dedicated QGIS plug-in to download a certain data set is available, then it can be downloaded manually from the Web and then imported as new layers into QGIS. We recommend you browse the pages of publicly available Earth observation products to learn more about them. This includes the product portals of the ESA Copernicus Programme, including https://scihub.copernicus.eu and https://land.copernicus.eu, the NASA Earth Observing System, including https://earthdata.nasa.gov and the USGS, including https://earthexplorer.usgs.gov.

3. Data preparation

In Chapter 2, we acquired a set of *vector* and *raster* layers covering the previously defined *area of interest* (AOI). These layers are still not ready for analysis since they represent different projections and have diverging extents. Thus, this chapter focuses on preparing newly acquired data for further analysis purposes.

3.1 Deciding on a projection

The Earth's geometric shape, referred to as *geoid*, can be approximated as a rotating sphere (e.g. a rotation ellipsoid). Any arbitrary point on the surface of such a sphere can be described by two angles, *latitude* and *longitude*, which form the geographical coordinate system. Latitude describes the angle between the perpendicular of the rotating sphere and its equatorial plane at a given point on the sphere's surface. Longitude describes the angle between the *prime meridian* and a plane through the North and South Poles at a given point on the sphere's surface. Unlike the equator, which is derived from the axis of rotation, the prime meridian is artificially defined as the meridian representing 0°, which separates the Earth into Eastern and Western hemispheres. This means that latitude and longitude indicate how far north, south, east or west a point on the surface is located. Transferring the three-dimensional geographical shape of the Earth's surface, or a subset of it, onto a two-dimensional plane, for example, to create a map that can be displayed digitally and printed, is called a *projection*; it represents a non-trivial mathematical problem that has been the objective of cartography for a long time. In contrast to the geographical coordinate system, a projected coordinate system is based on coordinates representing a plane instead of a sphere. However, converting one into the other cannot be done without adding certain distortions: a projection can only preserve a maximum of two of the four fundamental geometrical properties, namely area, direction, distance and shape. Due to this limitation, projections are often customized to preserve one or two of these properties. This is usually indicated by the name of the projection, for example, 'equal area', 'equidistant', 'azimuthal' (preserving direction) or 'conformal' (preserving shape). Additionally, there are three main projection types: cylindrical, conic and planar. All of these represent different mathematical approaches as to how the Earth's surface is projected onto a plane; thus, they have different advantages and disadvantages.

A projection is described by a *coordinate reference system* (CRS), which consists of two elements:

- The geodetic datum, which is a *geodetic reference system* placing reference locations on a geometrical body (e.g. a rotation ellipsoid), representing the Earth's approximate spherical shape.
- A *coordinate system*, which projects a coordinate grid (cylindrical, conic or planar) onto the geodetic reference system.

Every CRS uses specific spatial units, for example, degrees, feet or metres, depending on the geometrical properties it preserves as well as its projection type.

Different data providers deliver their data in different projections, which may have been selected by the provider depending on the product type or the region covered (local projections). Thus, it is highly probable that combining multiple spatial data products for a joint analysis will require the reprojection of at least some of these products or even all acquired products should none of the initial projections be suitable for the analysis.

Usually, selecting a suitable projection depends on the area covered by a study: for small-scale regional study areas, we recommend using a projection that best fits a reference ellipsoid representing the Earth's approximate spherical shape in a region to local reference points. Such local projections are of higher spatial precision in their *region of best fit* than global projections but are practically useless for other parts of the world. The region of best fit can range from small districts to a whole continent. In Europe, for example, the European Terrestrial Reference System 1989 (ETRS89) is used as a geodetic reference system by several projections to project any geographical location throughout the continent onto a two-dimensional plane. Similar, locally adjusted references exist for other regions, for example, the North American Datum of 1983 (NAD 83) for North America. Large-scale study areas with a world-spanning extent may be best represented by projections that use a reference ellipsoid fitted to a set of points globally distributed over the Earth's surface. This results in a locally less precise positioning of locations compared to local projections, but a global optimally reduced error in precision for any region. At the global scale, for example, the World Geodetic System 1984 (WGS84) is used as reference for many different projections.

Bridging the needs for a standardized global CRS and its spatial precision has been a longstanding issue in cartography. Projections designed to be precise in one place are impractical in another. Depending on local conditions, for example, geographical latitude or the aspect ratio of the area along latitude and longitude, the coordinate systems used by such projections also differ. This makes it difficult to work with spatial data covering a greater geographical extent than is meaningfully represented by local projections. With the development of the Universal Transverse Mercator (UTM) coordinate system, this issue has been resolved. The UTM system splits the Earth's surface into vertical, strip-type, 6-degree wide zones; each is separately projected onto a Cartesian coordinate grid using a transverse Mercator projection. UTM usually uses WGS84 as the reference ellipsoid. In other words, UTM WGS84 can be described as a globally applicable CRS that consists of 60 independent regional projections, one per zone.

Thus, selecting a suitable projection depends on weighting the advantages and disadvantages of the CRS available in the study area: very local projections offer high geodetic accuracy but use coordinate systems that users in other parts of the world may not be familiar with. Often, such projections are oriented along administrative borders, for example, of countries, and to historically developed reference measurement locations. Global projections make sense if the study area spans the complete or major parts of the Earth. Globally standardized regional projections, such as UTM, combine regional accuracy with a globally evenly interpretable coordinate system; thus, they may often be the best choice. Therefore, we will use UTM WGS84 for the present project. The zone best fitting our study area is *zone 32N* (see Figure 1.46 or the QGIS zone preview).

3.2 Reprojecting raster and vector layers

Although all layers in the QGIS session are displayed in the correct place and overlay each other correctly, they are only reprojected *on the fly* by QGIS and may physically represent differing projections. The target CRS to which QGIS reprojects the layers of differing projections is defined in the project properties (Figure 3.1) and can be viewed and adapted:

```
Project > Properties ... > CRS
```

Make sure that the selected CRS reads *WGS 84/UTM zone 32N* with the *Authority ID EPSG:32632*.

Changing the projection of a layer consists of two different operations that need to be carefully distinguished:

- Assigning a CRS: this means that you declare the CRS of a layer *without* recalculating its coordinates! Assigning a new CRS should only be done if no CRS is defined and you have prior knowledge of the original CRS that is associated with the layer's coordinates. Assigning a new CRS is not recommended because coordinates are no longer interpreted correctly. There is no way back, unless you know the correct CRS associated with the layer's coordinates.
- Reprojecting a CRS: this means that you need to define a target CRS into which the layer's coordinates are to be reprojected. Geospatial software such as QGIS ship with many default CRS and transformation equations between those CRS; this allows the software to transform coordinates from one CRS into another.

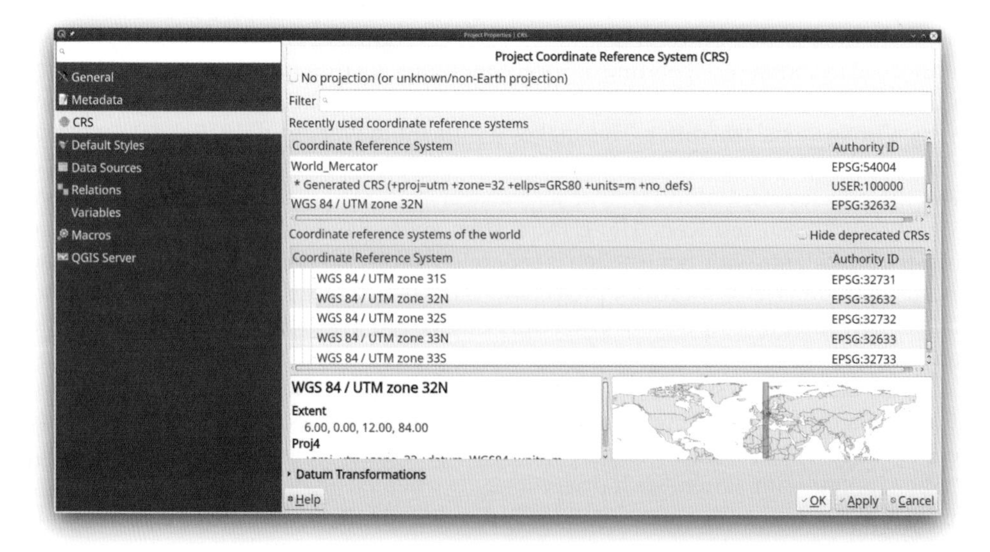

Figure 3.1 CRS settings in the Project Properties CRS window. For the selected CRS, all layers with differing projections are reprojected on the fly.

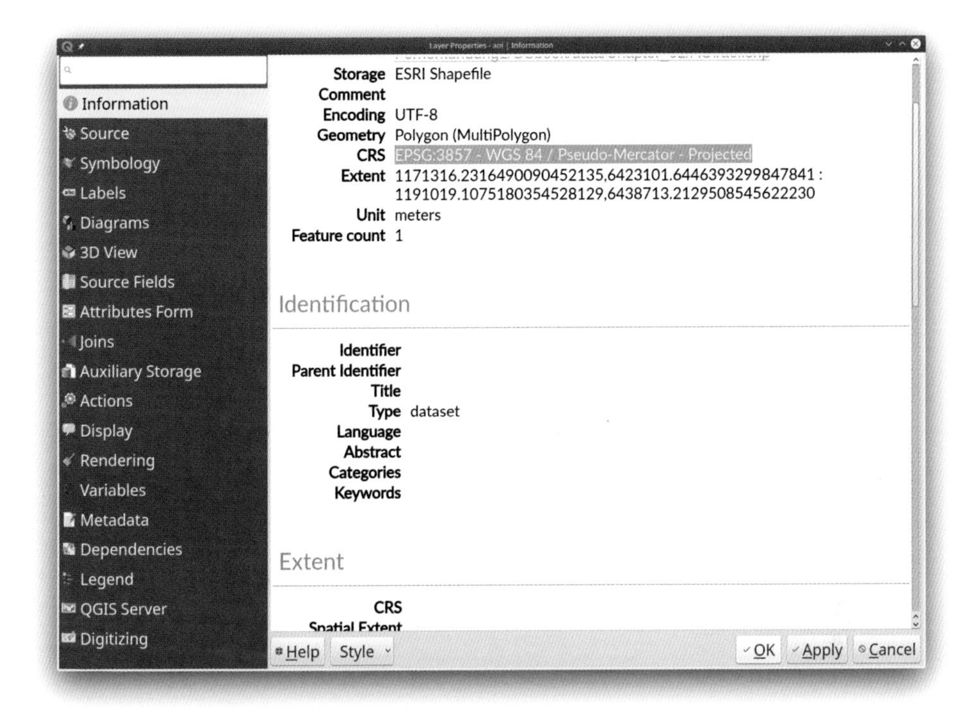

Figure 3.2 General information displayed in the Layer Properties window.

Let us first check the CRS of the AOI layer by right-clicking on the layer in the *Layers Panel* and selecting:

```
Properties > Information
```

In the *Information from provider* section, you can check the layer's CRS (Figure 3.2). The CRS of the AOI layer is defined as *EPSG:3857 – WGS 84/Pseudo-Mercator – Projected*, which means that the layer is projected.

TASK

Check out the CRS of all layers in the running QGIS session. Which layers are not projected (i.e. have a geographical coordinate system)? Which layers are projected, but not with the CRS we decided to use for this project? Which layers are already present in the CRS that we want to use?

Since we want all layers to be projected as *WGS 84 / UTM zone 32N*, you need to reproject those layers not present in the UTM. For the vector layers (AOI, place, admin_level, highway, landuse), select:

```
Vector > Data Management Tools > Reproject Layer ...
```

Select the layer you want to reproject (e.g. AOI) from the *Input layer* drop-down menu. Define the target CRS, which in our case is *WGS 84/UTM zone 32N* (the CRS of the raster

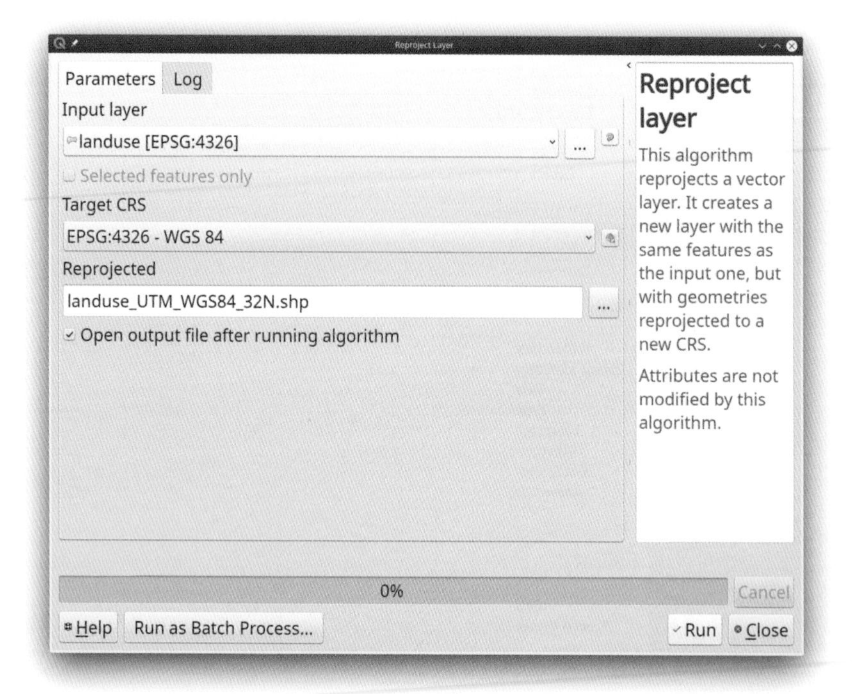

Figure 3.3 The vector layer reprojection tool in QGIS.

layers). Since we have already defined this CRS as our project CRS, it should show up in the *Target CRS* drop-down menu. If not, select the CRS symbol to the right of the menu and search for the CRS (e.g. by its *Authority ID EPSG:32632*). After selecting the target CRS, define where the output layer should be saved to and choose a name for it. We recommend naming the new layer so that you can easily differentiate it from its original layer, that is, for the layer named 'landuse' use 'landuse_UTM_WGS84_32N' (Figure 3.3). Select *run* to reproject the latter. It may be helpful to define in advance how files and layers should be named. While at the beginning of a project, naming conventions often seem unnecessary (e.g. because there are only a few files), the number of files will increase with each processing and analysis step. Thus, inconsistent naming across different steps can make it very hard to figure out what type of operation has been applied and what stage in the analysis a file represents. Therefore, we recommend using long, self-explanatory filenames than short, abbreviated ones that require you to remember what each abbreviation means. It may be convenient to add the operation used as the filename extension, for example, AOI_reproj_UTM_WGS84.shp, which should allow you to figure out all the steps already applied to a certain file just from its filename.

TASK

Reproject all vector layers as described here to the project CRS so that all layers inherit the same projection. Then, remove the original layers form the layer section since we will only need the layers sharing the same CRS.

INFO

File and layer names should be self-explanatory; specific modifications that have been applied should either be displayed within the filename or within an associated meta-information file.

Raster layers can also be reprojected by selecting:

```
Raster > Projections > Warp (Reproject)...
```

Since the Sentinel-2 raster layers of this project are already projected to *WGS84/UTM zone 32N*, only the SRTM DEM raster layer (*N49E010*) needs to be reprojected. As in this case, when working with raster data from different sources, they are most likely projected to different CRS.

3.3 Clipping to an AOI

Both vector and raster layers used in the project do not perfectly match our AOI. Instead, their extents extend beyond the AOI. Instead of applying our analysis to a much larger area then necessary, it is wise to clip all raster and vector layers to the AOI. Vector layers can be clipped using:

```
Vector > Geoprocessing Tools > Clip ...
```

Choose a vector layer as the input layer, for example, the reprojected 'landuse' layer (e.g. 'landuse_UTM_WGS84_32N'). As the overlay layer, select the reprojected AOI layer (e.g. named 'AOI_UTM_WGS84_32N'). Choose a filename indicating that the layer has been clipped, for example, 'landuse_UTM_WGS84_32N_clip'. Then, click on *Run* to clip the layer (Figure 3.4).

TASK

Clip all vector layers to the AOI layer as described here and name the output layers accordingly.

Like vector clipping, raster layers can be clipped to a certain extent or vector shape. Since raster layers with high spatial resolutions require a lot of computational power for some analytical steps, clipping raster layers to an AOI before analysis it can save a lot of time. QGIS offers two clipping tools for raster layers: *clipping by extent* and *clipping by mask*. For our project, it makes more sense to clip to a mask (our AOI):

```
Raster > Extraction > Clip Raster by Mask Layer ...
```

Select one of the reprojected Sentinel bands as the input layer and choose the reprojected AOI layer, to which you previously clipped the vector layers, as the *mask layer*. Use the defaults for the remaining options. Choose an output filename. The Sentinel band layers currently are saved as .jp2 files. To create data products that can be more easily read by

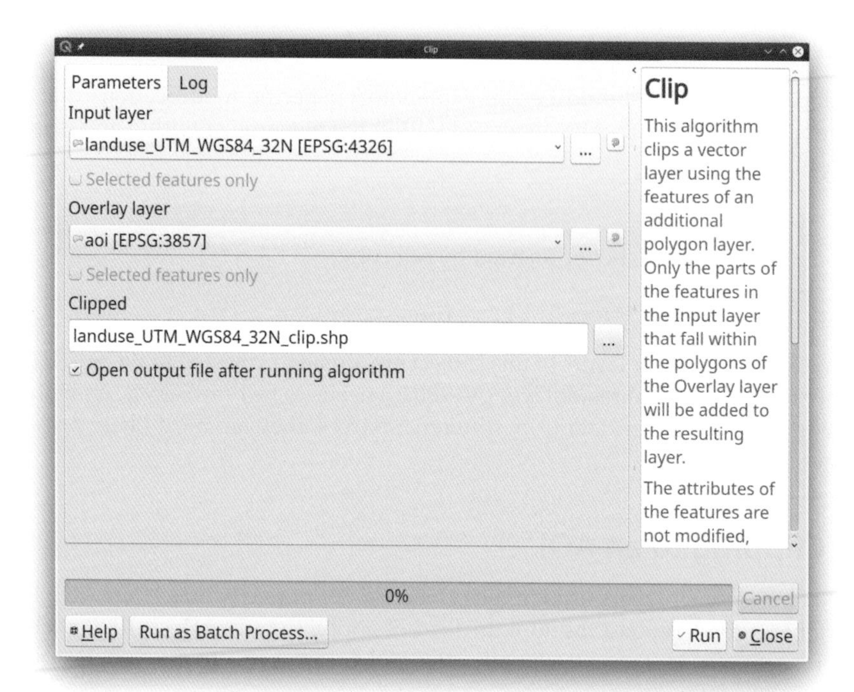

Figure 3.4 The vector layer clipping tool in QGIS.

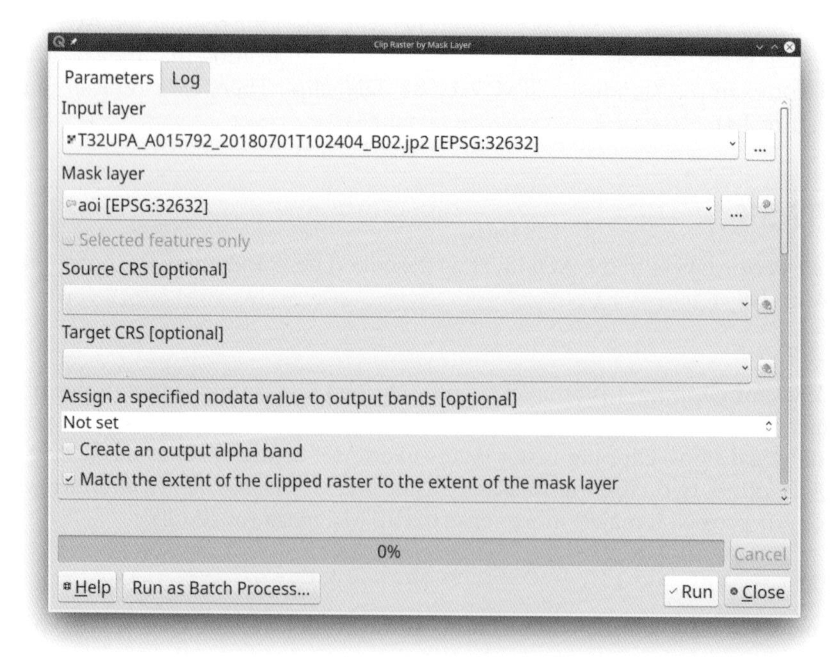

Figure 3.5 One of the two raster clipping tools in QGIS: raster clipping by mask.

other applications, it makes sense to use another file format. *GeoTIFF* is a widely used format for spatial raster files that can be read and interpreted by many applications. Thus, to use the GeoTIFF file format, add the .tiff extension to the filename. We recommend using this format for all raster outputs created within the following tasks. Click on *run* (Figure 3.5).

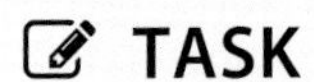 **TASK**

Clip all raster layers to the AOI layer as described here and name the output layers accordingly.

3.4 Stacking raster layers

Currently, each clipped Sentinel raster layer represents a single band. It can be advantageous to combine them into a single object, a process called *stacking*, for example, to visualize them properly or to carry out certain analyses, such as image classification. Raster stacks can be saved to a single file, which contains each spectral band. In QGIS, stacks can be created by building a virtual raster (VRT):

```
Raster > Miscellaneous > Build Virtual Raster ...
```

This will open the stacking tool (Figure 3.6). Select all Sentinel raster layers as the *input layers*. Since we only downloaded Sentinel-2 bands with a 10-m spatial resolution,

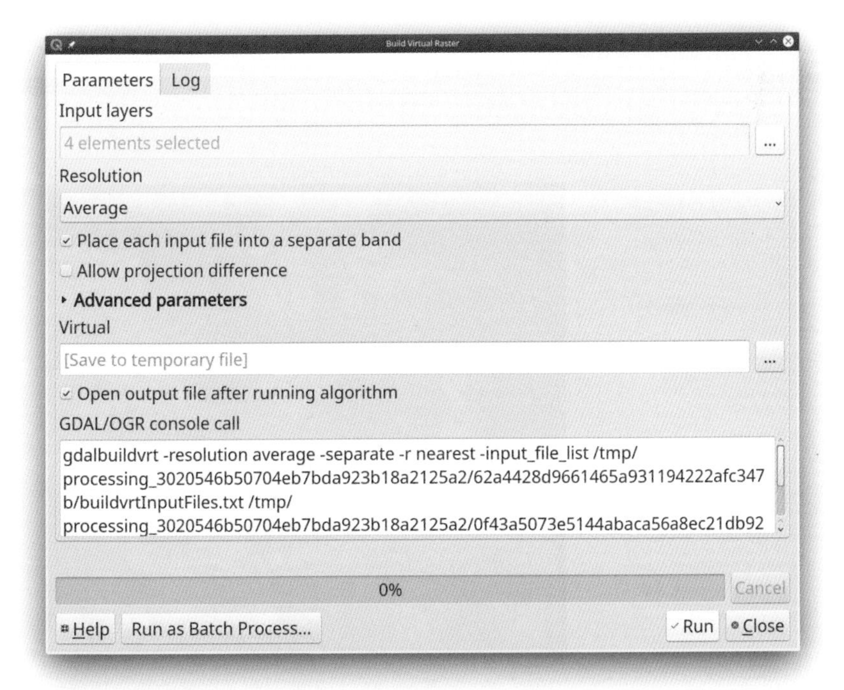

Figure 3.6 Building a VRT: single raster layers are stacked into one raster stack.

the setting *Resolution* is irrelevant. If you wanted to stack bands with different spatial resolutions, this setting would define the spatial resolution of the entire raster stack. Tick the option *Place each input file into a separate band*. For now, we do not need to write the VRT stack to a file since we will export the stack as a GeoTIFF file after it has been created. Thus, do not define an output file. The VRT will instead be saved as a temporary layer and will appear in the *Layers Panel*. Click on *run* to start the stacking.

The newly created VRT stack is now shown in the *Layers Panel*. Currently, this is a temporary file that virtually combines four bands, which are still stored independently. For our analysis, we will need to combine these layers into one file. This can be achieved by exporting the VRT layer; right-click on the layer and select it from the drop-down menu:

```
Export > Save As ...
```

Choose a location and filename ending with the suffix .tiff to save the stack as a multiband GeoTIFF stack. Load the GeoTIFF stack back into QGIS. You can now delete all previous single-band raster layers from the *Layers Panel* since all the raster processing steps have now resulted in a single raster stack, which is used in the following chapters.

3.5 Visualizing a raster stack as RGB

Even though the newly loaded raster stack may look like a normal RGB image, it is displayed in pseudocolours. The reason for this is the default band order QGIS uses to display a stack, which is the opposite of the band order of the raster stack. Thus, blue values are displayed as red and vice versa. To change this, double-click on the layer in the *Layers Panel* and select the *Symbology* register. Choose 'Band 3' for 'Red band', 'Band

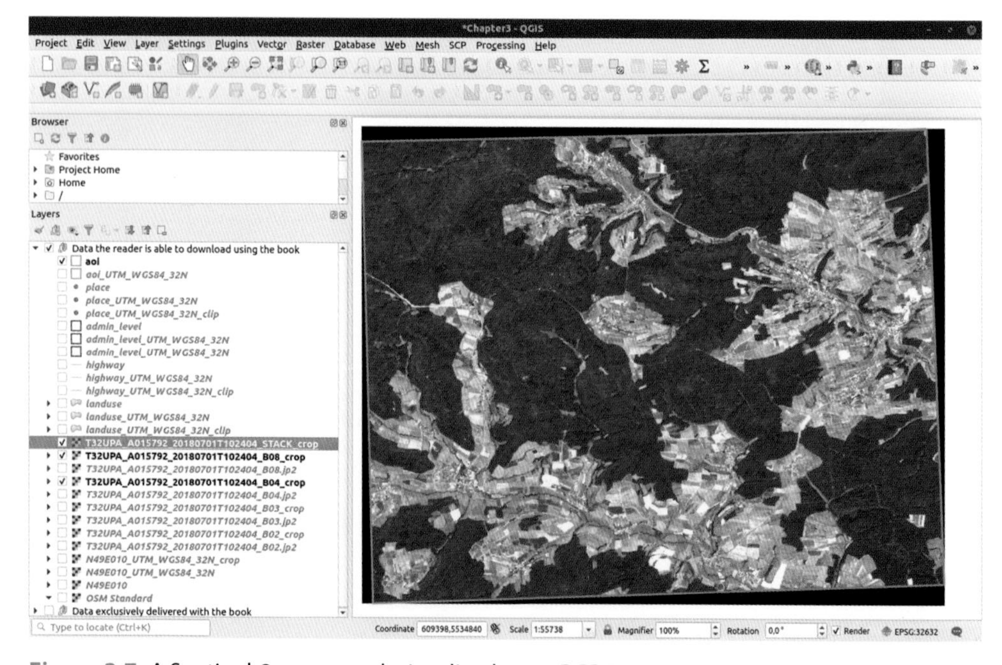

Figure 3.7 A Sentinel-2 raster stack visualized as an RGB image.

2' for 'Green band' and 'Band 1' for 'Blue band' and click on *OK*. Now, the bands of the visible domain are displayed as RGB (Figure 3.7).

3.6 Summary and further reading

These processing steps are common data preparation actions that are important because they must be considered and applied within each project dealing with spatial data from different sources. In summary, defining a consistent file naming structure not to lose track when differentiating intermediate data products from different processing steps is crucial to conduct an analysis efficiently. First, we recommend to reproject all data sets to a common project CRS to be sure that they overlap each other, and their extents share a common AOI. Once this is done, it is safe to clip all layers to the AOI and stack layers that belong together or are needed as a single physical file for later analysis. The effects of certain projections on how geoinformation is displayed is very difficult to understand without visualizing certain projections side by side. If you are interested in exploring geographical projections by comparing them visually, we recommend you have a look at the Compare Map Projections website (https://map-projections.net/compare.php).

4. Creating maps

Maps are created for different purposes: as data overview for personal use; for orientation during fieldwork; or to present the results of an analysis. All these kinds of maps have their rationale. They require several key map features (Figure 4.1) but differ to some extent regarding information or design elements.

Creating maps is usually done when analysing the final data to then present it; it is nearly as important as the analysis itself. Being able to present or communicate your results appropriately is the key to making your work understandable so that people grasp the impact and relevance of your work. Providing your target audience with a bad map might lead to your work being underrated or encourage negative feedback. Therefore, it is important to invest time to convert your results into understandable and user-friendly maps. Beside some fundamental principles, individual adaptation based on different messages or the background of the audience should be considered.

A map to be used in the field can be created without any design elements; however, at the very least, it needs coordinate labels for the axes and a grid. It should also display important landmarks, cities/towns/villages, the road network and your field sites, thus making it easier for you or your colleagues to navigate in the field and find the field sites. A map for presenting you results should focus on delivering such key information. In this book, the key information is usually a spatial data set that represents the results of an analysis. However, the audience may also need additional information for orientation purposes, such as cities or national park borders. Finally, a map to be used in a presentation needs to be visually appealing. An amateurish map might undermine the professional scientific analysis you want to deliver.

A *fieldwork map* should consist of:

- A coordinate system.
- A coordinate grid.
- Coordinate axis labels.
- A scale bar.
- An arrow indicating north, if the map is not oriented northwards.
- All kinds of landmarks required for orientation.
- Road and rail network.
- Cities and villages (with names).
- Any environmental information you need to see during your fieldwork (e.g. land cover, elevation).
- Your field sites or any locations you want to sample.

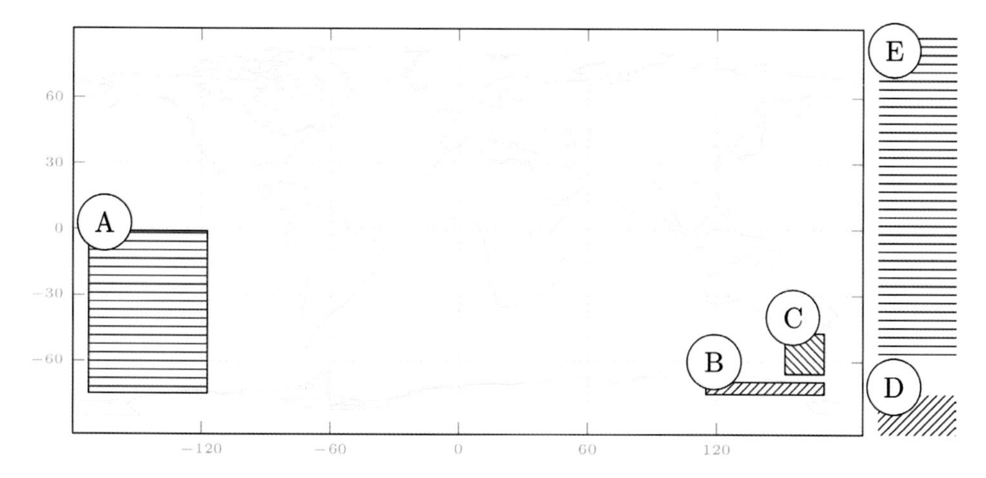

Figure 4.1 Commonly, beside the actual spatial data, a map provides a legend (A), a scale bar (B) and an arrow indicating north (C). Key attributes of the spatial data, such as the source of the data, the date when the data were acquired and the author are also needed (D). Additionally, information about the results, the author and institution can be added (E).

A *results map* should consist of:

- A coordinate system.
- A coordinate grid.
- Coordinate axis labels.
- A scale bar, an arrow indicating north.
- Only key landmarks, cities and roads required to understand the area.
- An overview map to orientate the user to where the study area is located (if needed).
- The key results of your analysis (based on raster or vector data).

Creating good maps, especially those used to present your results to the public, may take days to weeks to create and should not be done in a rush. A (good) map showcases your key results – if the audience does not understand your research because of a poorly designed map, then all your hard work would have been wasted! Thus, your map should look uncluttered while providing all the necessary information and have a pleasing design that reflects your professional work.

Especially for maps depicting large-scale studies, for example, continental or global study areas, the impact of the projection and its attributes must be considered. For example, the distance might not always be the same across the whole study area (Figure 4.2). Additionally, distortions might be present in your maps that are due to a persistent issue with projecting a spherical surface onto a plane: distance, angle and area size can never be preserved at the same time (Figure 4.3). For small-scale studies, you can usually ignore such distortions. However, if it is important that one of these spatial features is not distorted to visualize your result correctly, you should check the characteristics of the distortion(s) of your selected map projection and whether it is adequate for the specific message you want to transmit via your map. You should also consider changing the projection for the purpose of visualization.

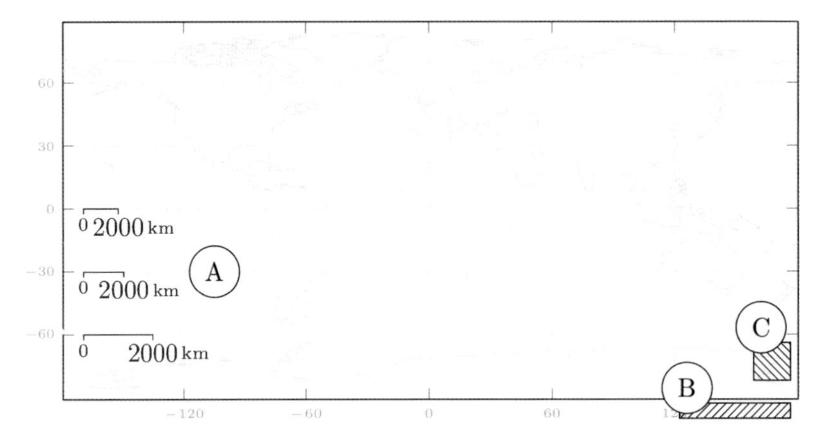

Figure 4.2 Map scales are crucial to understand the extent of the study area; however, depending on the extent of your study area and the projection used, the scale might differ across your map (A). General information about the data and its source is also required (B). An arrow indicating north (C) is especially important if your map is not oriented northwards.

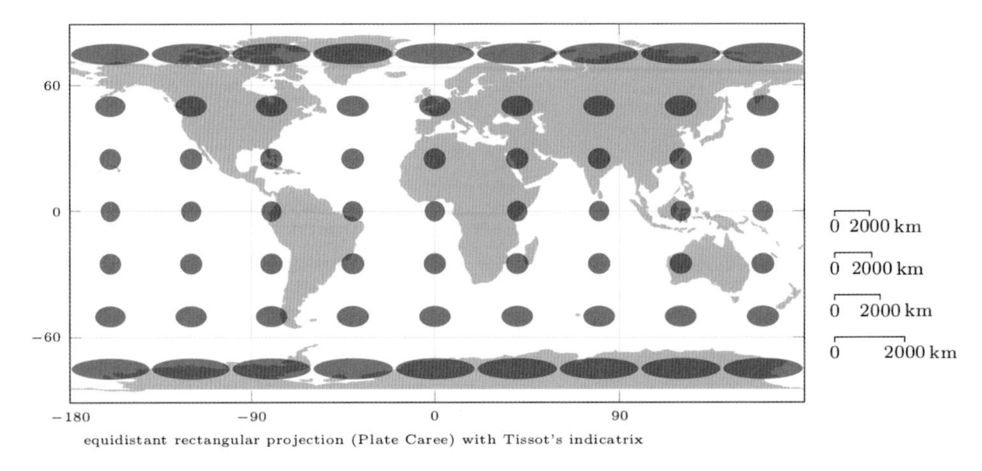

Figure 4.3 The impact of projections on shape and distance must be considered for large-scale or global maps. Projections do not preserve angle, size and distance simultaneously.

> **R TIP**
>
> Plotting maps in R and including a scale bar will prompt a warning message if your distance across the map extent differs. Hence, the length of the scale bar cannot be applied across the whole landscape.

In certain cases, providing an overview map to guide the reader is recommended (Figure 4.4). This is especially true for small-scale studies where you cannot assume that everybody knows the location. Overview maps should be straightforward and provide

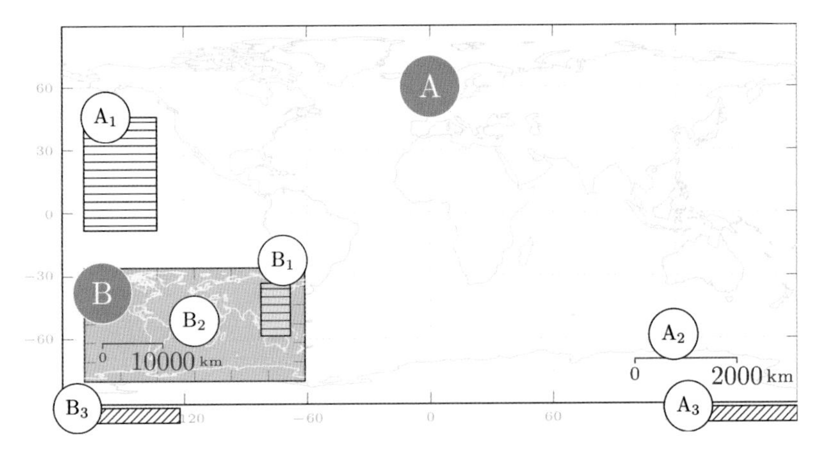

Figure 4.4 Overview maps (B1–B3) within the map of the study area (A) can play an important role if the overall location of the study area is not commonly known. Both map types require separate labels (A1, B1), scale bars (A2, B2) and information sections (A3, B3).

a general idea as to where your study area is located. Thus, shorelines, national borders and major cities could be added to the overview map, including a rectangle representing where your study area is located. The actual study area map provides the results along with some key landmarks or cities for orientation purposes.

4.1 Maps in QGIS

Creating maps in QGIS is quite straightforward and can be done quickly; however, improving the design might be time-consuming and should be included in your planning.

First, we will create a simple map for fieldwork; then. We will then create a map to be used for presentation or publication purposes. For the following map examples, we use the data from Chapters 2 and 3. After working through the next sections, you can reopen your current map layout project and add your final spatial results to your map.

The QGIS map manager or *print layout* can be opened within QGIS via:

```
Project > New Print Layout ...
```

If you have already created a map layout, you can select the *Layout Manager* or click on a layout name in the *Layouts* interface, all within the *Project* drop-down menu. The new layout interface has a main window dedicated to map design and several property tabs on the right, which are dedicated to the general layout or specific items we will add in the next steps (Figure 4.5). Saving or creating a new layout, as well as exporting it, to insert it into a presentation can be done via various functions on the *Layout* toolbar. Zooming in and out or zooming to the full extent of all items in our map layout window can be done using various functions found on the *Navigation* toolbar (Figure 4.6). More operations can be done within the *Toolbox* toolbar, such as adding items to your layout, like the map currently displayed in the main QGIS window. Items can also be selected, which allows the modification of their properties within the *Item Properties* panel on the right-hand side of the window. An important issue with the existing *pan* and *zoom* options is that we can either apply to the item itself or to the map within the item. The *Move Item* and *Move Item Content* in Figure 4.7 refer to this issue. Changing the zoom factor of the item content,

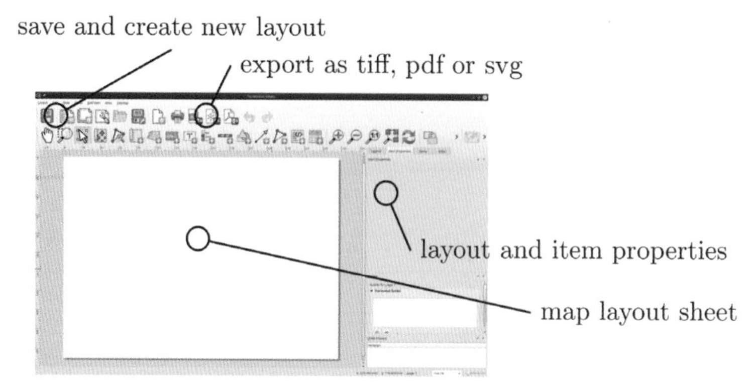

Figure 4.5 QGIS print layout view for creating a fieldwork map.

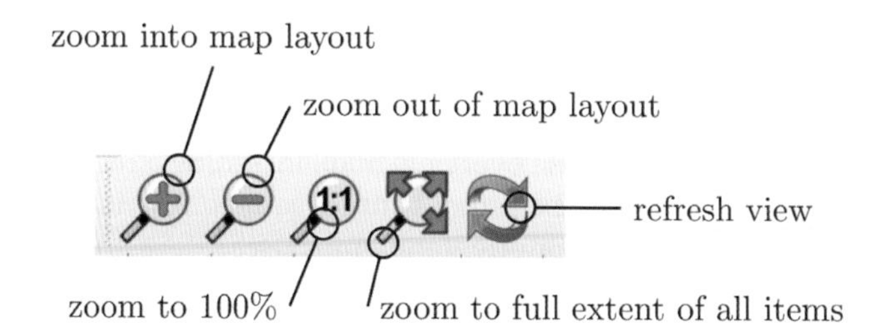

Figure 4.6 Tools that can be used to position a map in the QGIS Print Layout view.

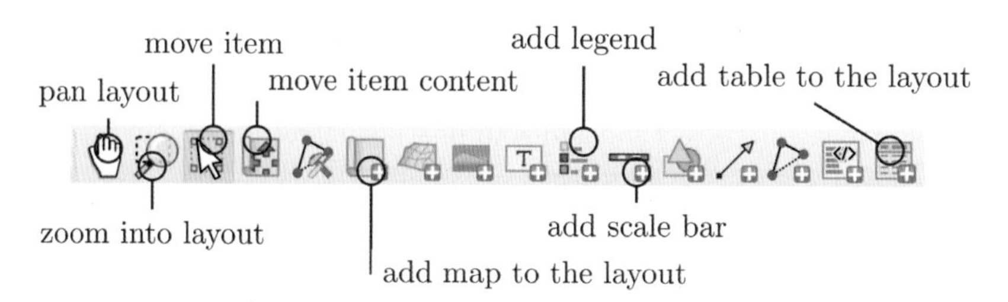

Figure 4.7 Tools for navigating and adding elements to a map in the QGIS Print Layout view.

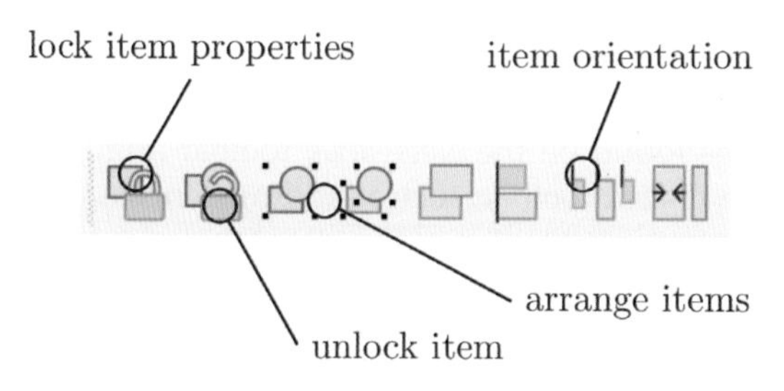

Figure 4.8 Tools that can be used to edit map elements in the QGIS Print Layout view.

namely the map, is also possible when the *Move Item Content* button is activated and the position of the mouse wheel is changed. Holding Ctrl while moving the mouse wheel makes for smoother zooming.

The *Map Action* toolbar allows us to lock an item; such an item will not be changed by any modifications made in the main QGIS window, such as through zooming or changing the layout (Figure 4.8). All items can be arranged and aligned using drawing software such as *raising items, grouping items* or *aligning items left*. Selecting an item and clicking on *delete* on your keyboard will delete this item.

TASK

Zoom and pan through the map layout sheet and then move and zoom only the content of a map item.

Adding a map to the layout via the button in the *Toolbox* toolbar (Figure 4.7) assumes that you have already selected the data you want to display as well as its layout and the desired extent within the main QGIS interface. Text formatting, such as a shadowed text, buffered text, text colour or font type, as well as the colour of the raster data set, must be defined in the *Layer Styling* panel (right-click on any toolbar or panel and select the *Styling* panel) within the main QGIS window. It is quite common to have both windows open side by side or to move between these two windows until all settings are properly defined (Figure 4.9).

For the fieldwork map, you should include a raster, the settlement layer and the road network. The raster should be a true colour composite, so that we can interpret the landscape. Settlements should be displayed as points with names; however, the road network should be more informative because we need to know where we can drive. Thus, we change the road network in the main QGIS window to *categorized* and *column*

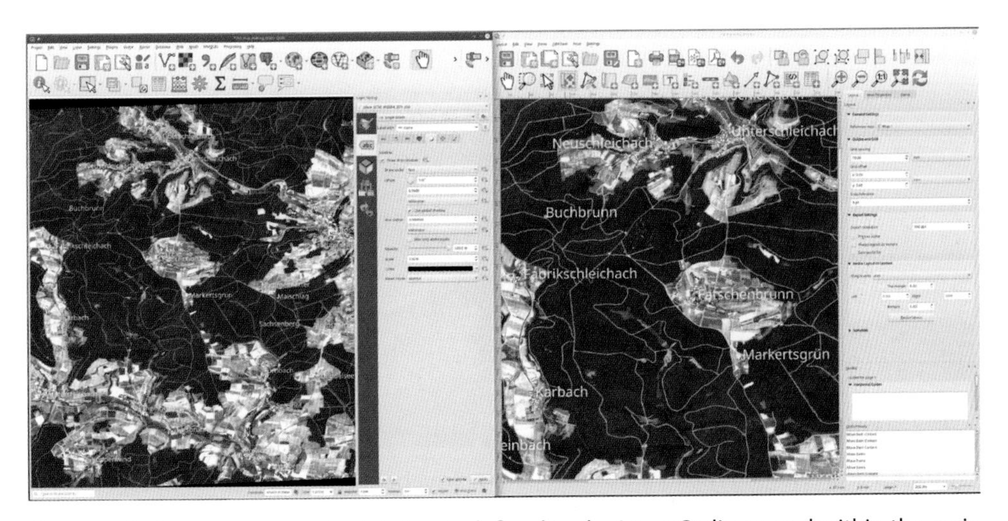

Figure 4.9 Raster, vector or text styling is defined in the Layer Styling panel within the main QGIS window. It will consequently impact the way objects are displayed in the mapping interface. Therefore, working on both interfaces for certain tasks is needed.

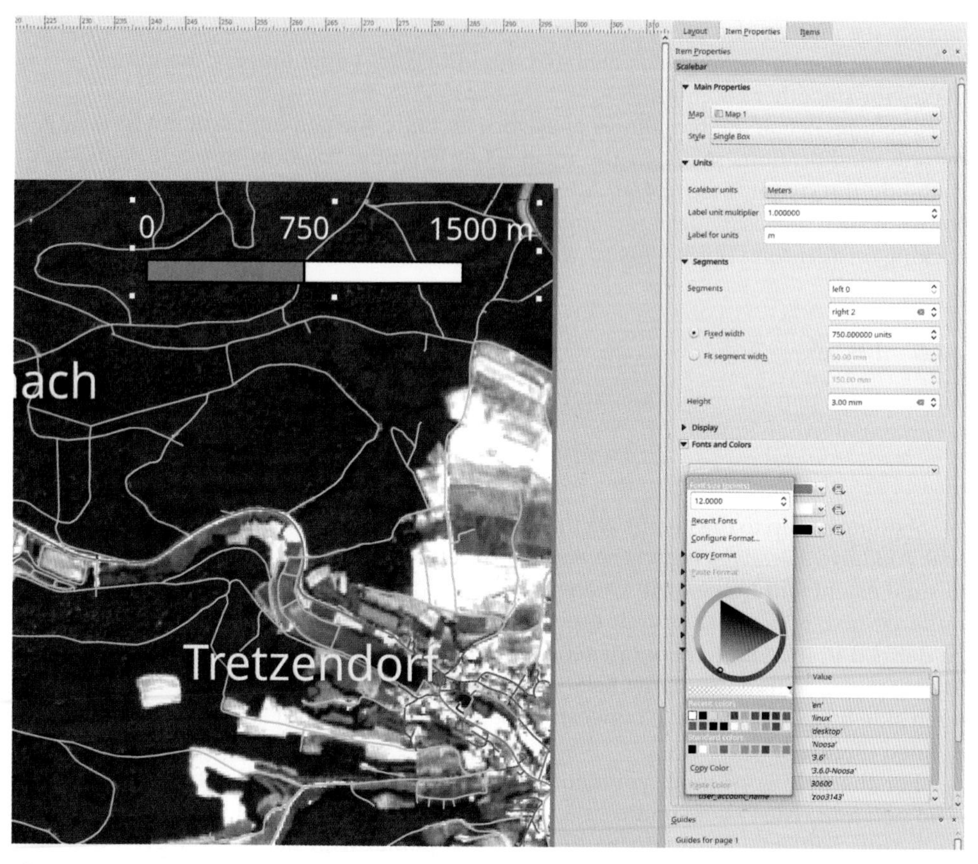

Figure 4.10 Within the Item Properties panel, the colouring of the scale bar can be defined by clicking on the down arrow. Further functionality is provided by clicking on the font button.

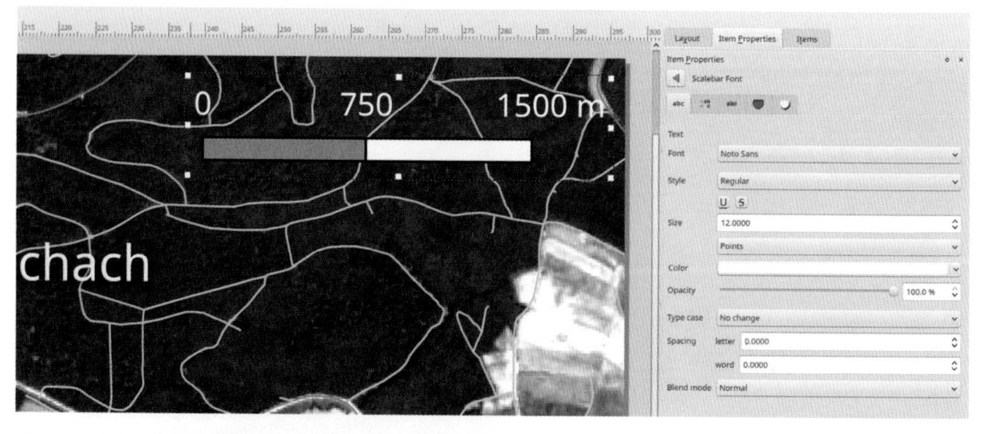

Figure 4.11 Clicking on the font button opens a Scalebar Font interface that offers various options on how to style your scale bar. Text colour, size or type can be defined in this interface. We set the colour to white because of the dark background; however, shadows can also be inserted to increase the contrast between text and background.

place. By clicking on *Classify*, we get different symbols for the existing categories. These are also used for the map layout.

Within our map layout, we add a scale bar and position the item on the top left corner. The values are scarcely legible; thus, we need to change the font colour via the *Item Properties* panel. After expanding the *Fonts and colors* tab, we can click on the down arrow on the *fonts* tab (Figure 4.10). By clicking on the actual *Font …* button, we get a more extensive property window (Figure 4.11). Each option opens different styling interfaces that can be used to customize text appearance. After setting the text colour to white, we can also customize the scale bar colours, for example, to blue and white. Additionally, we can add a legend so that you or your colleagues can understand the items on the map. After clicking on the *new legend* button, we can position a new legend using a mouse left-click and drag a rectangle on the preferred area. All items in the *Layer* panel of the main QGIS window are listed; however, usually not all items need to be displayed in a legend. Within the *Item Properties* panel on the right, we can see a *Legend Items* window if we select the legend item on the map. Deselecting *Auto update* activates the buttons below, which can be used to remove, add or rename items. Additionally, font size and spacing within the legend can be set (Figure 4.12). After changing the item properties, all items can be moved around on the map and positioned correctly. Locking an item ensures that it will not be moved or changed accidentally.

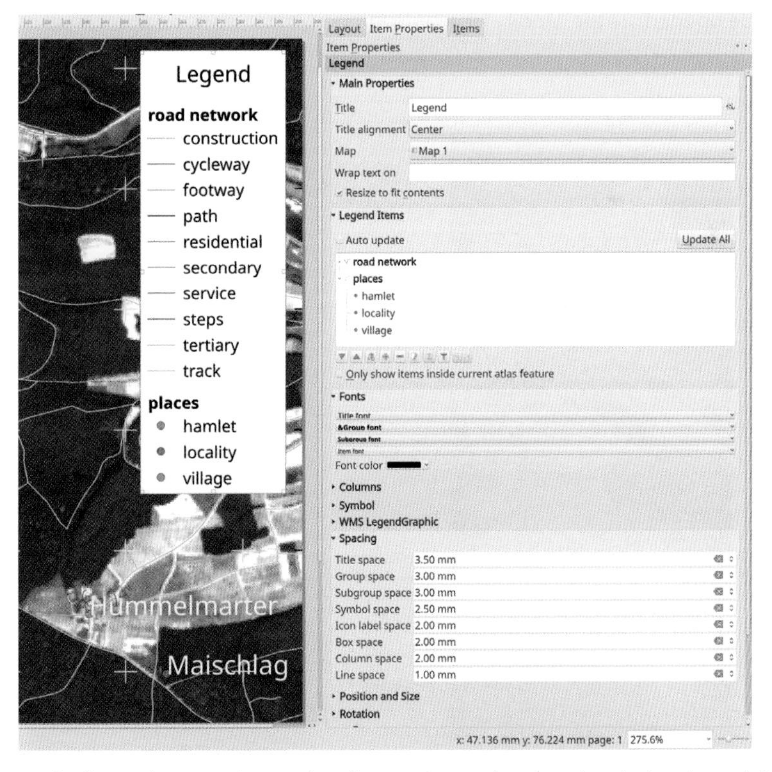

Figure 4.12 The legend properties can be changed once Auto update is deselected. Removing or adding entries can be achieved via the plus or minus sign. The sheet button can be used to change the name of an object in the legend. Additionally, the font size, colour or spacing between entries can be modified.

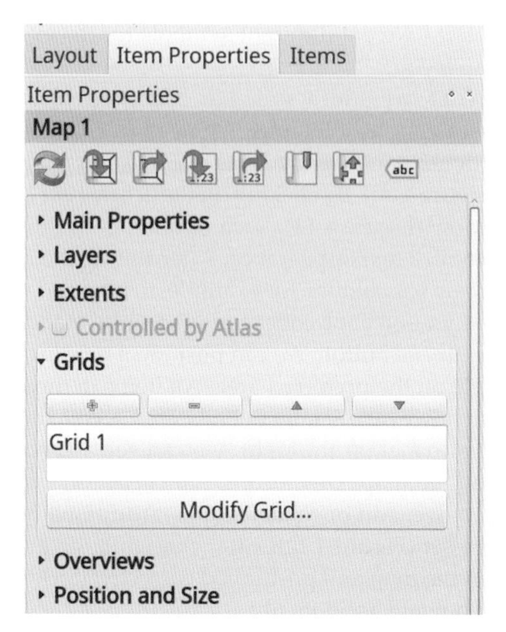

Figure 4.13 Within the Item Properties panel, the Grids option can be used to activate a coordinate system grid. Its styling can be altered in the Modify Grid panel shown in Figure 4.14, which can be opened via the Modify Grid button.

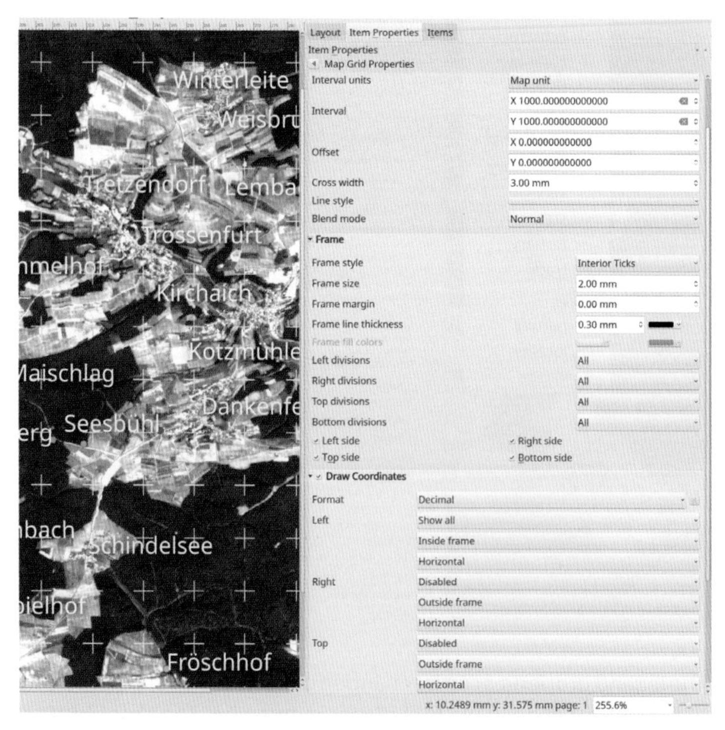

Figure 4.14 The grid properties define the styling of the grid. The interval between grid lines, as well as the frame style and the plotting of coordinates, are determined using this interface.

One last relevant visual guide on a map are the coordinates, so that your location and the coordinates provided by a global positioning system can be matched. After selecting the map, the *Item Properties* panel shows various options for map display. Among other properties, a grid can be added through the *Grids* section (Figure 4.13). Once added, the grid can be modified. Colour and type (cross, marker, solid), as well as the interval between grid lines, frame style and where to draw the coordinates along the axes can be defined (Figure 4.14). In our example, we set the type to *cross*, the interval to *1,000* for the x- and y-axes of the grid lines and *interior ticks* for the frame style. Other frame styles and intervals or colours can be explored. The *Draw Coordinates* checkbox can be used to define where the coordinates should be displayed. We chose to draw coordinates on the left and bottom, but not on the top and right map axes. Also, we defined the coordinate values to be plotted inside the map and made them *vertical ascending* on the left axis to align the values with the map border. Once this was done, we created a simple map for fieldwork that provides us with the key information regarding settlements, road networks, distance and coordinates (Figure 4.15). In this simple example, some information that is relevant for colleagues, for example, which coordinate system has been used, is still missing. Usually, it is good to note who created the map, when the map was created and which data (or data version) were used. The latter is especially relevant for a constantly changing data set, such as new study sites and their names being added. Knowing which file version has been used allows the user to identify potential problems, such as wrong names having been incorrectly plotted in older file versions or where newer spatial files, including further study sites, may already exist.

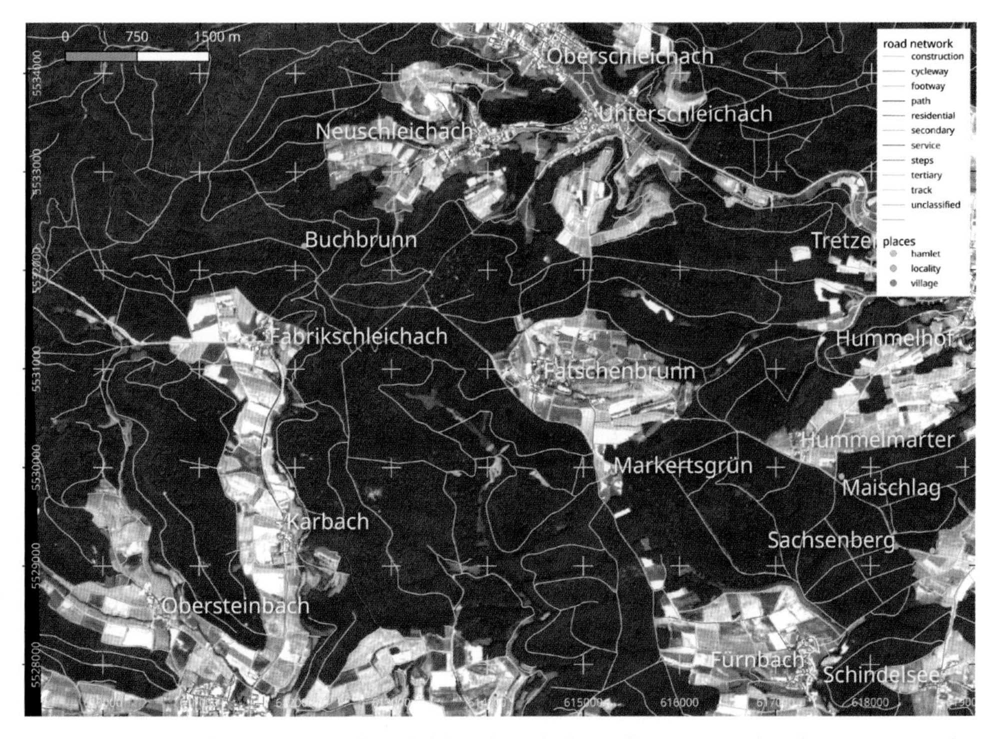

Figure 4.15 Simple map to conduct fieldwork, including all necessary details to navigate the study area and collect in situ data. The coordinate grid and scale bar are especially relevant for the navigation. The roads and settlements are useful to work out the best path towards a certain area of the study site.

> **TASK**
>
> Display a raster stack and change the RGB image to true colour (how Earth would look like from space) and false colour (using spectral values in RGB not visible to the human eye). Change the band assignments to the colour settings and interpret the result.

4.2 Maps for presentations

A map to be used for a presentation, poster or publication should be less practical compared to a fieldwork map. Instead, it should provide the key messages from your research findings and be visually appealing. Depending on the type of presentation, certain information must be provided. For example, for oral presentations, most of the details will be explained by you, whereas on a poster or in a publication more details must be provided on the map, such as its purpose, an overview map or the source of the data. A simple example of a map that could be used to present your study area is shown in Figure 4.19. This map clearly introduces the location within Europe, Germany and the local settlements. The background data (OpenStreetMap stamen data via the *OpenLayers* plug-in) is of no specific interest, whereas the geographical position of the rectangles and the names in the two detailed maps are.

> **TASK**
>
> Change the grid coordinates to be inside or outside the frame; change the style of the frame and the actual grid from lines to cross or point styles.

The overviews map and the connecting lines can be added to the map layout using the *Overviews* option (Figure 4.16). First, at least two maps need to be added to your map layout, one as the main background and the other one on top of it. Via the *Items* panel, items can be arranged so that they are on top of other items or align with them. After selecting the main map where the overview rectangles should be displayed, the *Overviews* option within the *Item Properties* panel can be accessed; a new overview can be added by clicking on the green *plus* sign. Now, the map frame that defines the position of the overview and extent can be added. Should you forget which map item has which name, just click on an item and check the name above the *Item Properties*, then select the main map again, return to the overview section and select the map you want to use. Once you have done this, you can define the style of your overview rectangle. Different overview rectangles can be added to the same map using different styles (Figure 4.17). Connecting the lines between the overview rectangle and the corresponding map can be added by using the *Add Arrow* button. Changing the style to 'End maker: None' in the *Line Markers* section, and changing the desired colour, width or line type within the *Main Properties* will result in lines connecting our overview extent and map (Figure 4.18).

Adding maps, changing the data to be displayed, adapting the style and making further modifications are shown in the example map shown in Figure 4.19. Once you have obtained your results, adapting the map you are going to use for presentations simply

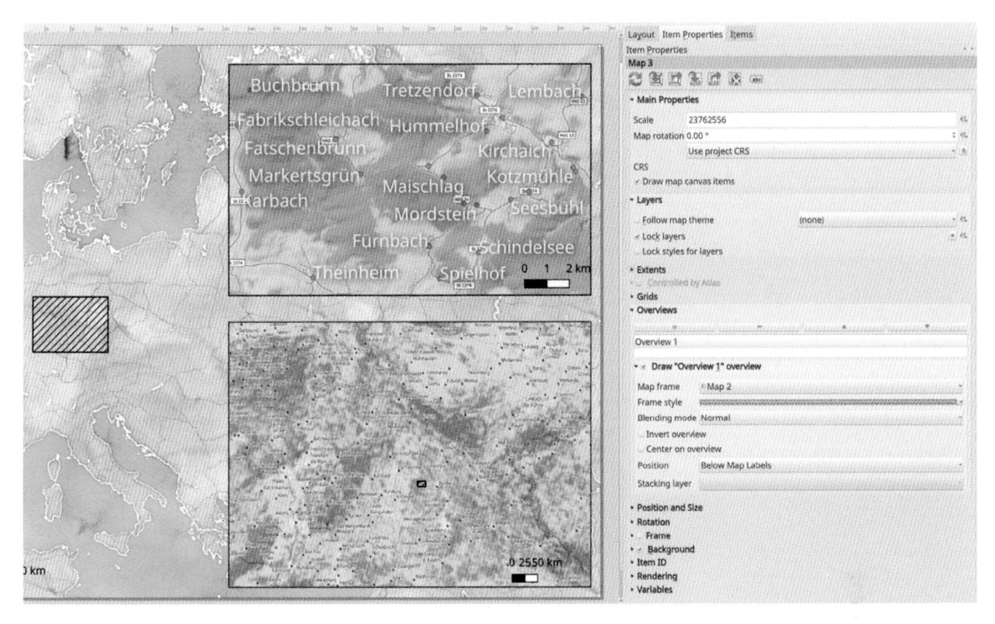

Figure 4.16 The geographical extent of maps can be included in another map. The extent and positions will change according to the extent and position shown in the defined map.

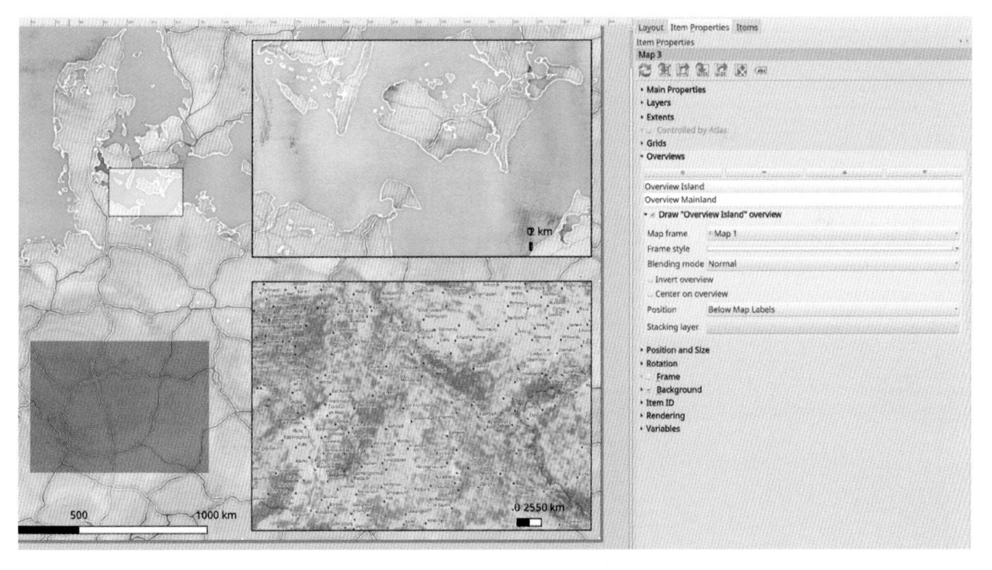

Figure 4.17 Various overview rectangles can be added and correspond to the geographical position and extent of the assigned second or third map. To obtain a better overview, the names and style of the overviews can also be changed.

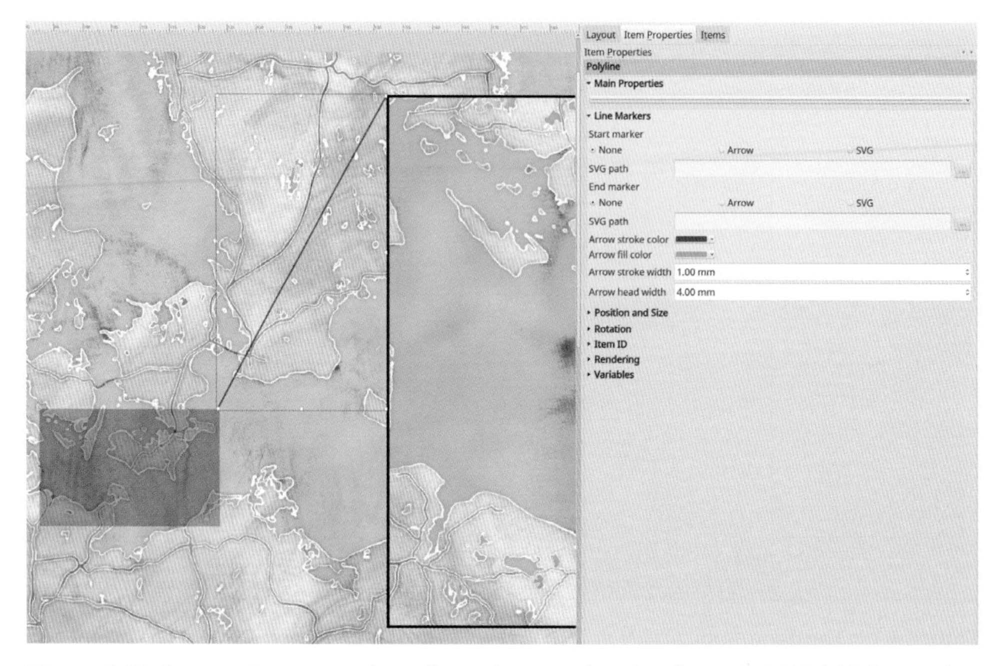

Figure 4.18 An overview rectangle and map item can be visually connected with lines using the Add Arrow tool and adapted with the line style.

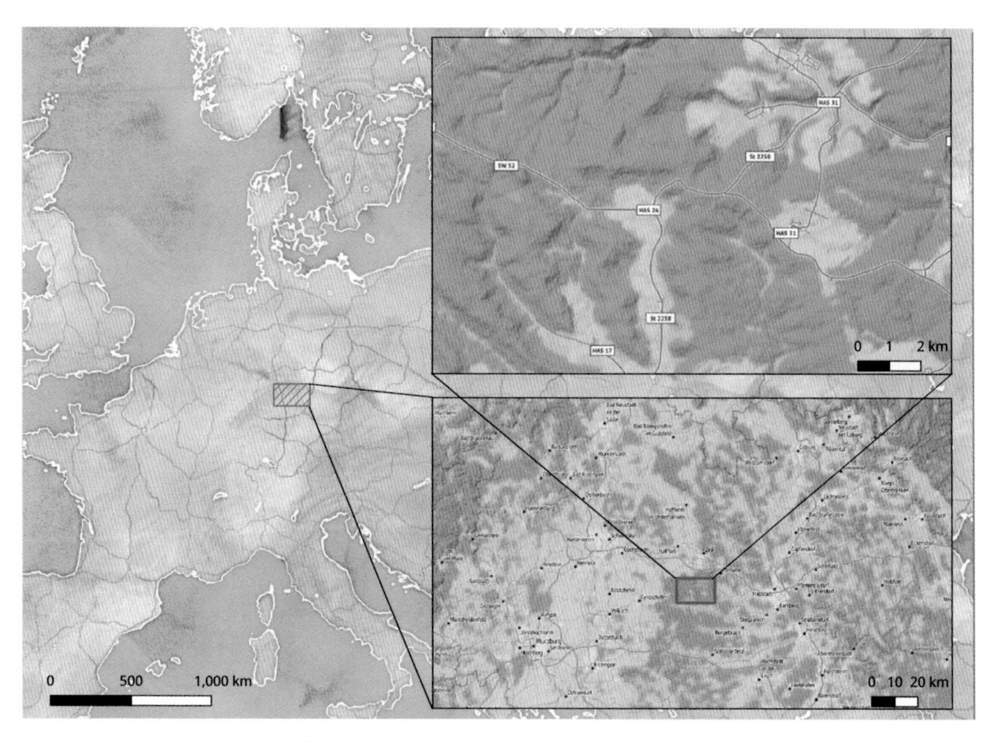

Figure 4.19 A map to introduce the study area with two overview maps and the corresponding extents depicted as rectangles. Background layers are OpenStreetMap stamen watercolour; the terrain is accessed through the OpenLayers plug-in.

requires the activation of that specific data set in your main QGIS window. Refreshing the specific map should display that data. If you want to prevent the displayed data from changing, remember to apply *Lock layers* in the *Item Properties* panel for each layer whose data you no longer want to change. Otherwise, the map will be updated using the displayed data currently in use in the main QGIS window.

TASK

Include an overview rectangle. Within the *overview section*, select the 'invert overview' option. Change the colour of the inverted overview to white and slightly translucent to highlight the area of interest in your overview map. Move the second map and change its extent. Does the overview map also change?

Another example of a map created purely for the purpose of a presentation or poster is shown in Figure 4.20. A road network vector file, for example, from OpenStreetMap, the Socioeconomic Data and Applications Center (https://sedac.ciesin.columbia.edu) or Natural Earth (http://naturalearthdata.com) can be downloaded and plotted on top of, for example, a shaded relief raster. To create a style with such density of colouring, the

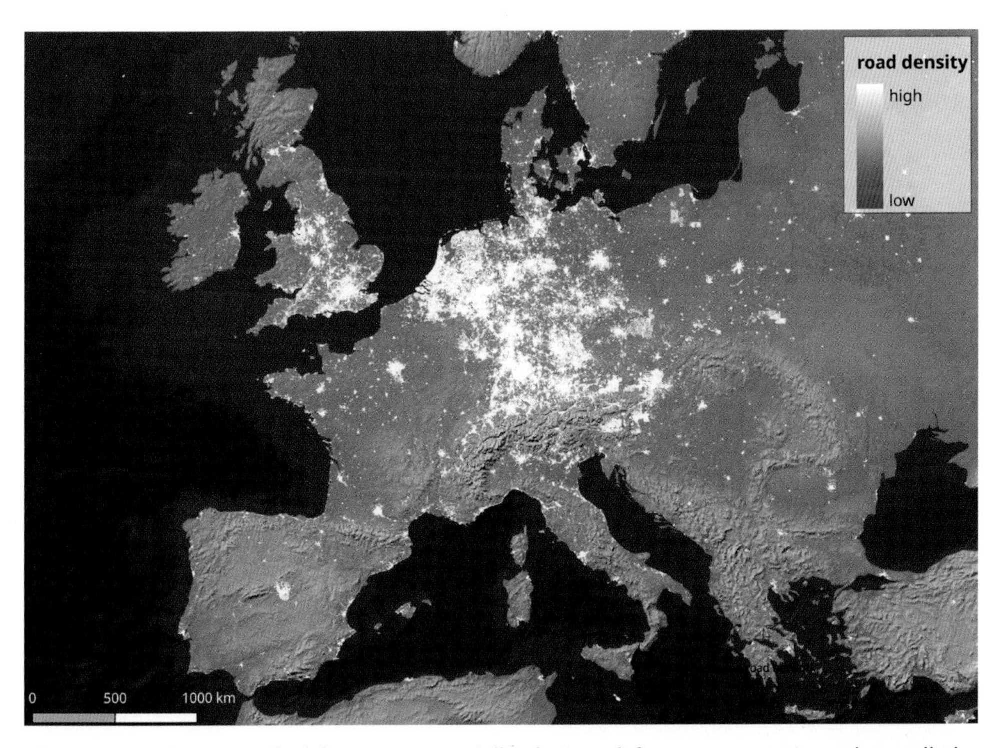

Figure 4.20 An example of a map especially designed for a presentation where all the necessary details are explained. Important features in this map are the source of the data used and its spatial completeness, as well as the way data are displayed because in this map the majority of Europe appears to have low road density. The idea for this map originated from the book QGIS Map Design by Anita Graser and Gretchen Peterson.

road vector lines are set to blue; within the same styling panel, at the bottom of the *Layer Rendering* section, the *Feature blending mode* should be set to 'Addition'. Settings allowing further changes in line width, colour or transparency allow you to fine tune your final map. The legend is created manually using the *Add shape* function to create a rectangle with a colour gradient from blue to white. The text and background have also been added manually. The shaded relief raster and the other layers can be accessed from Natural Earth or the Socioeconomic Data and Applications Center. Any other data set or data source, for example, ship or plane traffic data, could also be used within such map styling.

In some cases, this kind of map design can be very informative, but for many features (e.g. location of settlements) or additional information such as road length, this approach may not work. One approach used to generalize spatial information involves hexagons, as shown in Figure 4.21. This can be done with the same road network data set we used for Figure 4.20. First, we create a hexagon grid for Europe:

```
Vector > Research Tools > Create Grid
```

Within the *Create Grid* interface, we define the type (hexagon (polygon)), extent and horizontal and vertical spacing. For our example, we chose 200 km; for other areas or data sets, a different spacing might be more appropriate. The size of the hexagons also impacts the subsequent statistical values and thus the information on your map. The new hexagon layer is added to your layers and is called *grid* by default, unless you have already saved it under a different name. Next, we want to assign the road length behind each hexagon to the hexagon grid using:

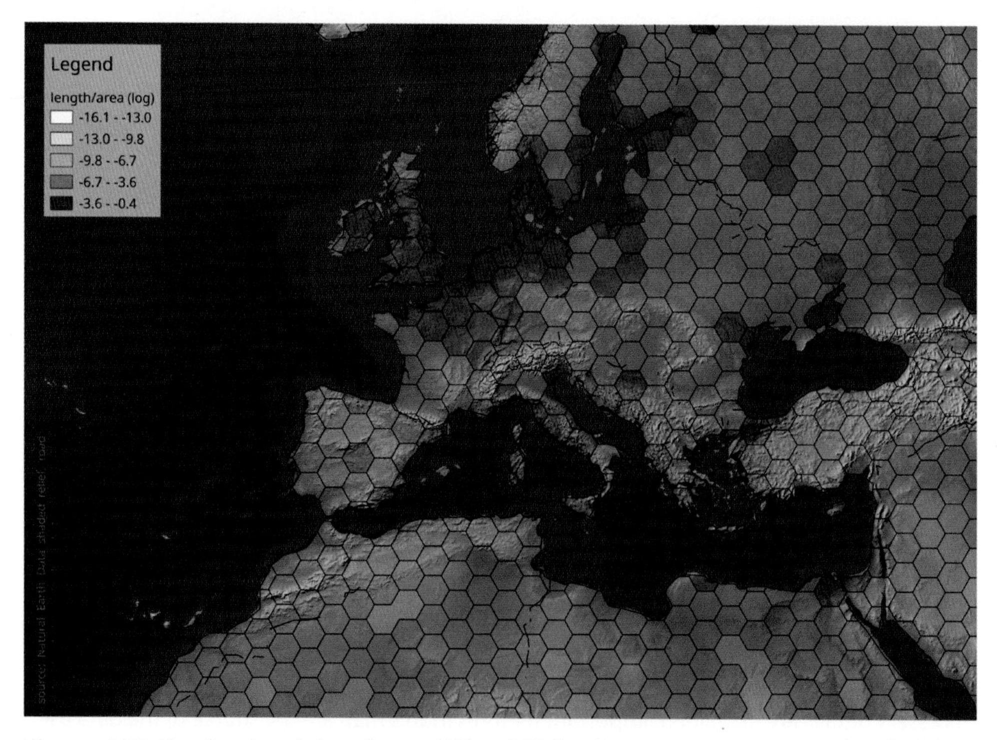

Figure 4.21 Road network length per 200 × 200 km hexagons as an example of how to generalize spatial information when the original data or information is too fine and thus might not be visible on a map. Data from http://naturalearthdata.com.

```
Vector > Analysis Tools > Sum Line Lengths ...
```

Now, we have another layer called *line length* (default name), which we can use to colour our hexagons based on the length of the roads within each hexagon. We can do so using in the *graduated* style option in the *Layer Styling* panel, with an *Equal Interval* mode and a blue colour ramp. The styling can be changed, as well as the number of classes, to achieve the best informative visual display. However, hexagons are shown for the whole extent, not only for the terrestrial areas where we have roads for. We can set all hexagons with a value of 0 to *do not display* (by ticking the class legend in the *Layer Styling* panel), but some parts of the hexagons may also cover ocean areas. Clipping the shape of all countries with our hexagon map would result in having a hexagon vector object that only covers the countries but not the ocean:

```
Vector > Geoprocessing Tools > Clip ...
```

The input layer is the hexagon object; the overlay layer is the *NE Cultural Admin 0 Countries* (via QuickMapServices) layer or any other layer you want to clip to. The resulting object solely covers the terrestrial area of Europe, showing the road length per hexagon (Figure 4.21). As already mentioned, both data and methods impact the resulting map, for example, here all hexagons along the coast have a lower road length value due to larger aquatic areas being covered. The area of the hexagons and the clipped hexagons can be calculated within the *Geometry Tools > Add Geometry Attributes*; the ratio is displayed in the *Layer Styling* column section via the *expression function* (the icon to the right of the column field or by typing *'LENGTH' / 'area'* in the column field). This allows you to colour the hexagons depending on the ratio between road length and the size of the respective hexagon. In some cases, value ranges must be adapted to achieve a better visual display of the pattern. In our example, we have many very low values, hence the *Equal Interval* mode does not result in an appropriate map design. Either we change the mode to *Quantile* or we apply a logarithmic function in the expression function to stretch the data ranges for our visual mapping purpose (ln('length'/'area')).

TASK

Change the order of the processing steps. Does it impact the results? Correct for the size of the clipped hexagons.

QGIS INFO

Layer styles can be copied and pasted onto other layers by right-clicking on a layer > Styles > Copy ...

Many freely available maps, such as the shaded relief map, are perfect as backgrounds for large-scale maps but are too coarse for local-scale maps. Therefore, such maps must be created individually. Using our digital elevation model (DEM), we can produce various informative and visually appealing background maps (Figure 4.22). This map has been created using a DEM (e.g. Shuttle Radar Topography Mission (SRTM)) which we loaded into QGIS. In *Layer Styling*, we applied the *hillshade* style. It is an alternative to *Singleband*

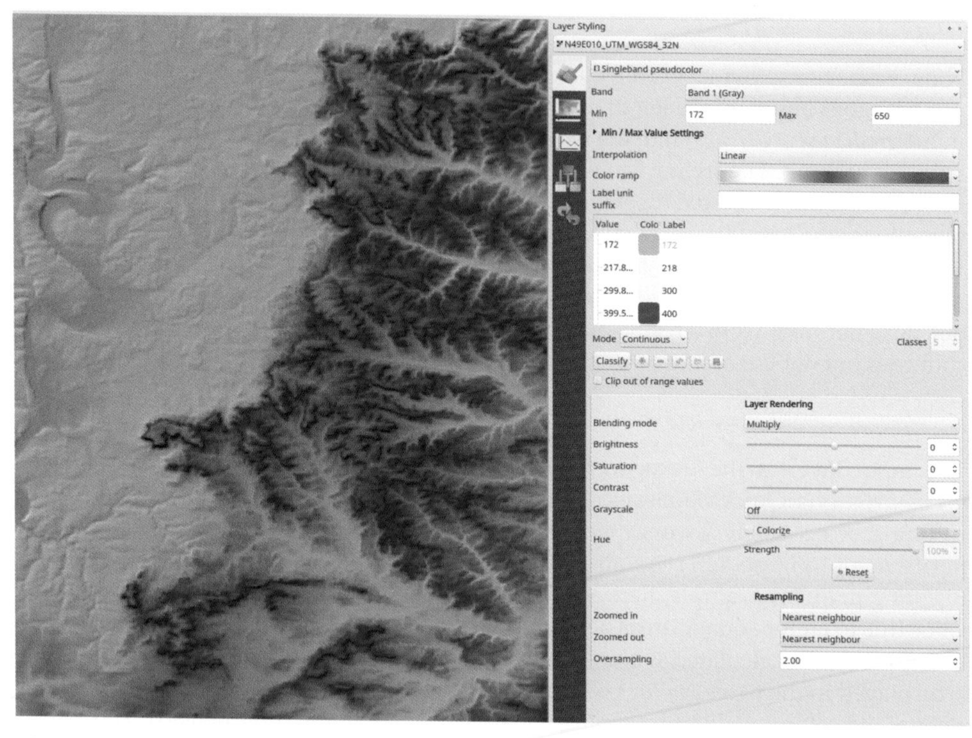

Figure 4.22 Hillshade map with coloured elevation values based on the downloaded SRTM DEM data set. DEM, digital elevation model; SRTM, Shuttle Radar Topography Mission.

gray or *Singleband pseudocolor*. Then, we copied the layer (by right-clicking in the *Layer* panel, *duplicate Layer*) and set it to *Singleband pseudocolor*. The colour ramp can be defined as desired; the important part is the *blending mode* within the *Layer Rendering* section, which must set to *Multiply*. This mode multiplies the luminance level of the current layer by the layers below; it maintains darker colours while removing white. The multiply mode is the opposite of the screen mode; the addition mode adds the pixel layer values, and the overlay mode combines the multiply and screen modes.

> **R INFO**
>
> Within R similar, maps can be created with the 'ggplot' and 'ggspatial' packages. Further packages allow to download spatial background data, such as 'rnaturalearth' or 'marmap'.

4.3 Maps with statistical information

Via the *Data Plotly* plug-in, QGIS offers a way to display spatial data with common statistical graphs. This is (not yet) available within the map interface but still offers some explorative mapping capability.

Loading data to explain its capability is feasible through the *QuickMapServices* (QMS) plug-in. After installation, an entry can be found in the *Web* menu; data such as country borders can be searched and downloaded:

```
Web > QuickMapServices > Search QMS
```

We selected the data set *NE Cultural Admin 0 Countries* and added it to our project (Figure 4.23). The map shows all countries; when the corresponding data table is opened, it lists all attributes available for the countries, such as *population estimates* or *income*.

After installing *Data Plotly*, the plug-in can be opened through:

```
Plugins > Data Plotly
```

Within the *Plotly* interface, plot type, as well as the layer, can be defined; which columns should be used to display on the axis can also be defined. Additionally, style properties can be defined (Figure 4.24). The unique feature of *Data Plotly* is that it allows you to display the values of a map on, for example, a scatterplot; by selecting single dots, the corresponding feature on the map is highlighted (Figure 4.25). Within *Data Plotly*, at the top of the graph window, various options such as *zoom* or *select* are accessible. Selecting a single dot in the graph highlights the corresponding country. Selecting a range of dots in the scatterplot using the *dotted rectangle* icon highlights all countries associated with the selected dots. *Data Plotly* graphs can be further customized with a legend, axes labels or by including a range slider using the tabs accessible to the left of the *Data Plotly* panel.

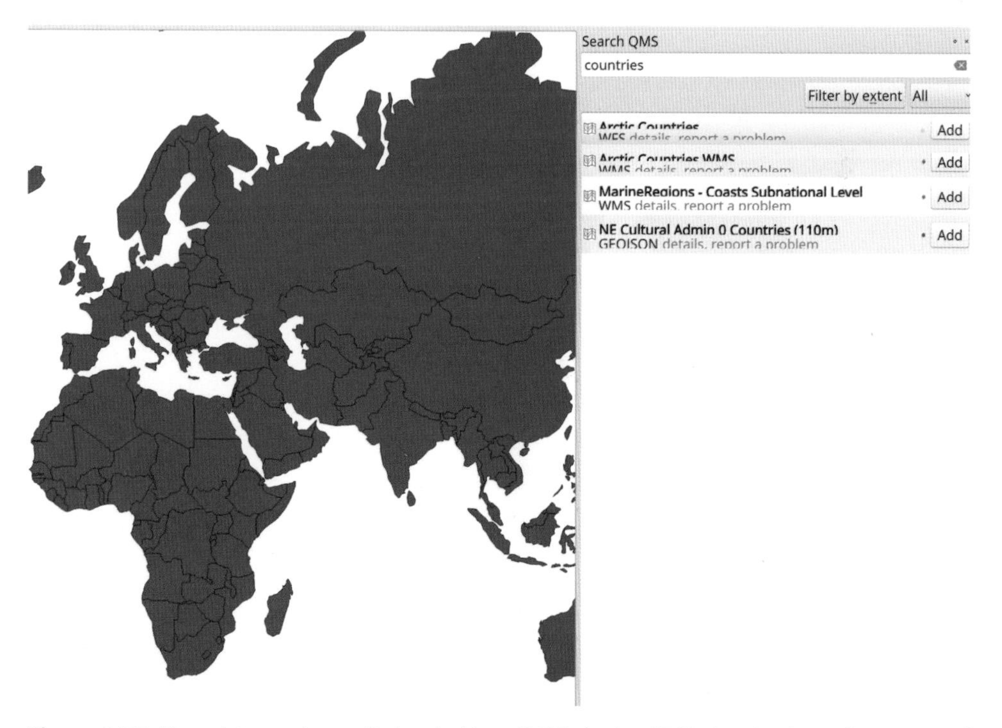

Figure 4.23 New data can be easily loaded into QGIS via the QMS plug-in. Apart from a set of default data, the plug-in also allows to search for further spatial data sets that can be added to your project.

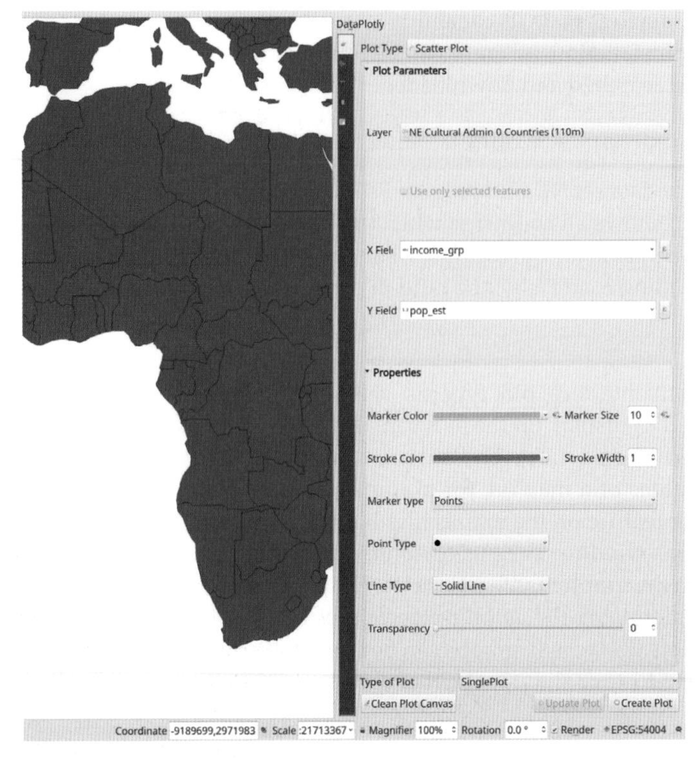

Figure 4.24 The data and type of graph to be used can be selected within the Data Plotly interface. Colours and the type of lines and plots can also be defined in the properties.

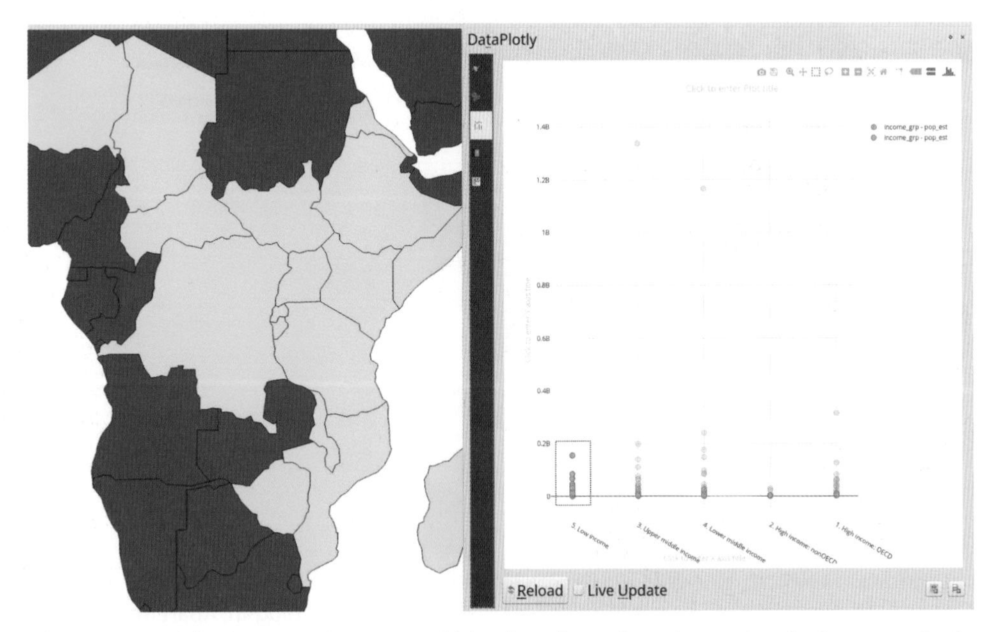

Figure 4.25 Selecting certain areas within the plot using the rectangle icon results in highlighted spatial features that correspond to the selected dots.

TASK

There are several ways to visualize map data. Try to use other plug-ins to create maps in QGIS. For example, investigate the animation functions within the *Time Manager* or *MMQGIS* plug-ins to visualize temporal changes.

4.4 Common mistakes and recommendations

When creating maps, a variety of common mistakes can be observed; these can be easily avoided. Like any other graphic, maps should be self-explanatory, meaning that all the necessary details must be provided. Never assume that the audience knows *where* your study area is or what *extent* your study area has; always include a scale bar, overview map or coordinates. A scale bar is especially crucial to understand landscape features. Knowing that we are looking at either a landscape extent of 1 × 1 km or 100 × 100 km is relevant; sometimes, the scale cannot be deduced from the depicted landscape alone. Moreover, your study area, region or country is not an island (unless it is one); therefore, the surrounding landscape or administrative boundaries should be shown on a map, perhaps using a muted colour or by adding a semi-transparent area on top of non-relevant areas, so that the focus remains on the study area. Especially if the map is rotated (due to a variety of reasons), an arrow indicating north is needed since we usually assume that a map is pointed northwards. Adding additional information, such as the source of data or the coordinate system used, depends on the type of presentation; details of the coordinate system used are needed for fieldwork and global maps, where scale, area or angle are important for your research.

If you create a map that is to be printed and distributed, it is important to add all this information plus the name of the map creator (you), any contact details, perhaps the logo of your institution and a short paragraph explaining the purpose and content of the map.

4.5 Summary and further reading

In this chapter, you learnt how to create maps using QGIS. The book *QGIS Map Design* by Anita Graser and Gretchen Peterson provides many great ideas to create visually appealing maps. We highly recommend exploring the various functionalities of QGIS map layouts. Making good maps is generally not related to which software you use; rather, it is governed by your visualization skills, your perception of proportions and colours and your ability to explore how your audience reacts towards different map designs. Further functions are available within QGIS, such as the *Atlas* function, which can be used to create a series of maps automatically to display different data or regions. Thus, if you are tasked with creating maps for 100 study sites, please consider this option within the *Map Layout* window of QGIS.

We want to emphasize that maps to be used for fieldwork should be tested beforehand and provided to colleagues who carry out the fieldwork with you to check their adequacy. The same is true for maps depicting your results. Considering any key messages and presenting and discussing drafts with your colleagues is important to create meaningful maps.

Part II

Spatial field data acquisition and auxiliary data

In Part II of the book, we address how to plan a field campaign, how to collect your own field data using a global positioning system (GPS) and how to get your data back into your geographic information system (GIS). We outline how locations can be transferred to your GPS as geographical positions and then be used to navigate in the field. Such locations are usually preselected field study sites, either based on prior knowledge or created using random or regular point pattern algorithms. We also describe how to export data from your GPS into your GIS and outline commonly performed initial analysis.

5. Field data planning and preparation

Collecting field data is a key aspect of most environmental studies and requires sound strategy and planning. In certain cases, study sites are already predefined, but often appropriate locations must be found before any fieldwork is carried out. In the following sections, how to select field sites and how to collect in situ data and integrate them into your geographic information system (GIS) are discussed.

A variety of questions need to be considered when planning a field campaign. These include:

- What is the scientific question?
- Which in situ data are needed to address the research question?
- Which environmental variables need to be recorded?
- Which technical equipment is required?
- Which area needs to be covered?
- Is it necessary to homogeneously sample in situ data for all land cover types within the complete area?
- Is the time of year important or can the data be collected in any season?
- Is one collection enough or do you need to collect data on the same location various times?
- Do restrictions for in situ collection apply (limited access, time constraints, etc.)?

Information collected in the field must be carefully reviewed. Before planning a sampling strategy in detail, it should be clear which environmental variables need to be collected and to which degree the accuracy of the samples is relevant. These decisions directly relate to the question of how time-consuming the sampling effort will be. Depending on your research and available time, either quantity or quality might be more important. For example, collecting data with very accurate measurements at five in situ sites situated within a very heterogeneous landscape, puts the whole analysis at risk since measurements would not reflect landscape variability. This problem arises irrespective of statistical issues with a sample size of five. Sometimes, you may also want to measure the same information several times in the same area, for example, a field or forest patch. This approach might be good to capture temporal variability at a site. However, such repetitive measurements within the same area must be regarded as one (n=1) sample when analysing them statistically; thus, they must not be used to increase the overall sample size.

Field survey date	staff name	
	date	
	location	

Point Measures	
name of field site	
GPS waypoint ID	
Lat	
Lon	

environmental attributes					
Landcover type					
attribute A	☐ *Dense*	☐ *Sparse*	☐ *Open*	☐ *None*	
attribute B	☐ *High*	☐ *Medium*	☐ *Low*	☐ *None*	
.....	☐ *Compact*	☐ *Linear*	☐ *Fragmented*	*Min-Diameter [m]* ☐	

Photographs / image names		
N 5 6 7 8	A (GPS)	5
	B (Plot)	6
	1 (_N)	7
	2 (_E)	8
	3 (_S)	
	4 (_W)	

Figure 5.1 The field data collection sheet needs to be designed carefully to reflect all scientific requirements and be practical to work with. Recording coordinates on the sheet and on the GPS is recommended. GPS, global positioning system.

A field data collection sheet (Figure 5.1), as well as a field data map, should be created that reflect the environmental variables to be collected as well as the sampling strategy. For example, taking multiple measurements over time for the same study site might require adjusting the collection sheet by adding a field for the name of the site and a field for the number of the repetitive measurement so that the data can be grouped afterwards. In certain cases, it may be beneficial to take pictures of the study site from various angles and link them with the spatial locations. This allows rechecking the landscape afterwards, should some questions arise, or comparing field measurements of various staff members by looking at these pictures. This option is provided by the eVis plug-in functionality within QGIS or by importing geotagged pictures if the camera stores the position in the meta file of the image.

Generally, we highly recommend that you check if your field data sampling and data collection strategy can easily be integrated into a GIS and in subsequent analytical steps. Just test your strategy outside your office, collect the data, import them into your GIS and run the desired analysis on the field data. Evaluate whether data collection is adequate, too few or too many details are collected, the data format is appropriate and how long the data collection takes.

5.1 Field sampling strategies

Setting your field collection sites should be as objective as possible and reflect the scientific question. Sampling can be done in various ways, for example, random, stratified random, regular or clustered (Figure 5.2). Random sampling does not take care of the potential oversampling of abundant land cover types. For example, random sampling might lead

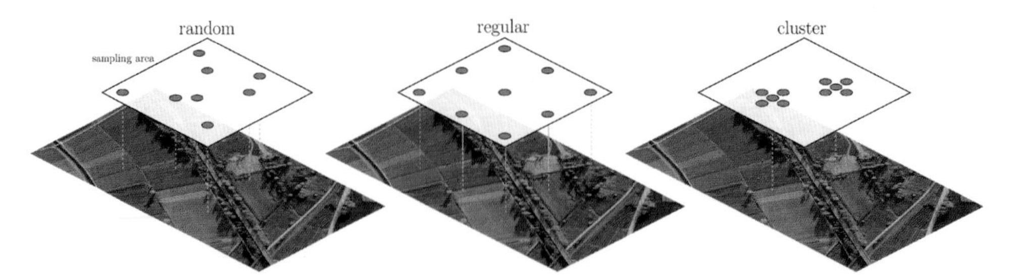

Figure 5.2 Different sampling approaches depending on the research aims and landscape structure.

to most samples being in a forested area whereas very few meadows are sampled, purely because you have more forest than meadows in the landscape. A stratified approach allows you to sample the same amount of locations in each landscape type. The definition of landscape types suitable for describing the sampling area need to be set beforehand and must reflect the scientific question. A corresponding spatial object that can be used to create the field data map needs to be available.

Thus, sampling can be done randomly across the whole landscape if certain land cover types are over-represented or under-represented or adjusted to certain landscape properties. The final approach very much depends on the characteristics of your landscape and your research approach. If the whole study region is covered by forest and no further differentiation of forest types is aimed for, a random approach is appropriate. However, if the forest or landscape varies across the study area, for example, due to forest types, elevation gradients or strong differences in climatic conditions, a stratified approach might be more appropriate.

5.1.1 Random sampling

Simple random sampling is the most common approach and ensures spatial independence. Within the vector menu, random points can be created either within an extent, the shape of another data set or specific polygons:

```
Vector > Research Tools > Random Points in ... (Extent, Layer
Bounds, Polygons)
```

Within either of these random point functions, you can set the number of random points and the minimum distance between points. In Figure 5.3, we used the 'landuse_UTM_WGS84_32N_clip' vector file and defined that 200 points with a minimum distance of 200 m should be sampled within the layer boundaries. Single polygons could also be selected manually. Within the random point interface, you can specify that random points should be created only within the selected features.

Additionally, areas or polygons, for example, representing specific land cover types, could be selected randomly to avoid personal bias towards one area or type. This can be done in QGIS via:

```
Vector > Research Tools > Random Selection ...
```

The number or percentage of features to be selected can be chosen. Using the

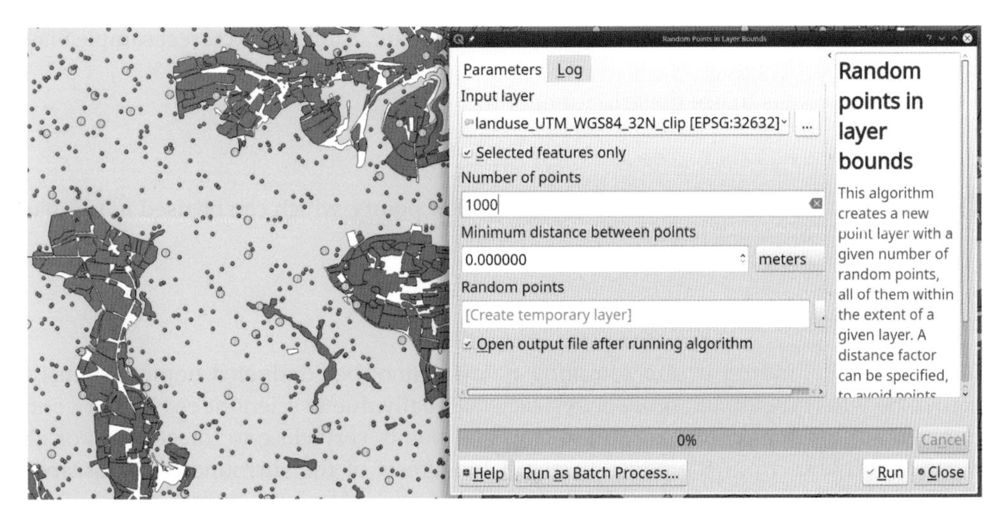

Figure 5.3 Interface and result of random points distributed across the whole layer boundary, with a minimum distance of 200 m between points (large circles) and 1,000 random points without minimum distance within the selected polygon.

landuse_UTM_WGS84_32N_clip vector polygon data set will result in polygons displayed in yellow (equivalent to being selected manually). The selected features can now be exported to a new vector file:

```
Layer menu > Right-click on layer name > Export > Save
Selected Features As ...
```

The newly created polygon vector file can now be used to either create a map for fieldwork or add sample points within these randomly selected polygons. Other types of

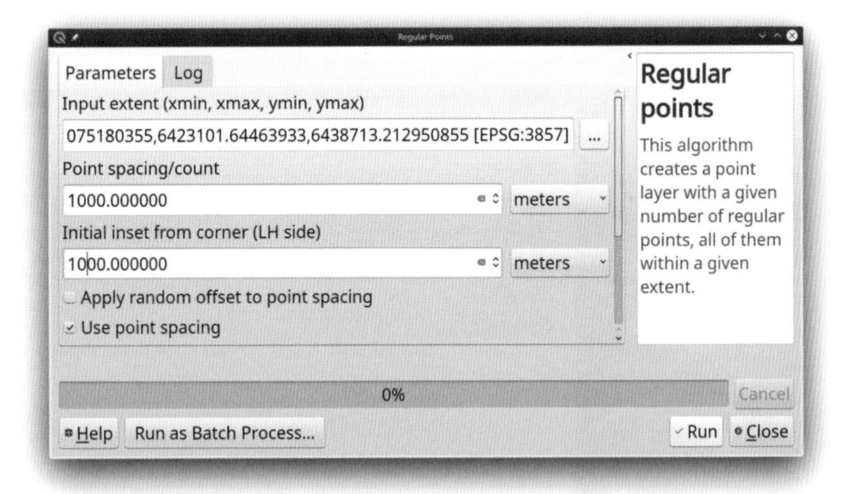

Figure 5.4 A rectangular point grid can be created and used to sample points regularly across the whole study area.

such selected sampling approaches are often applied, for example, to correct sample bias or answer a specific research question.

An alternative is to use the grid functions:

```
Vector > Research Tools > Regular Points ...
```

This function allows you to create a regular grid of points, which can be used to sample environmental attributes (Figure 5.4).

5.1.2 Selected sampling

In many cases, locating in situ sampling points cannot be conducted homogeneously across the whole region but must be restricted spatially due to scientific reasons. If your research approach requires you to set sampling points very close to forest edges or at a certain distance from the edges to avoid the effects of the surrounding landscape, you need to adapt the sampling methods accordingly. Moreover, certain areas might not be accessible, for example, due to potential threats such as mines. Because of time

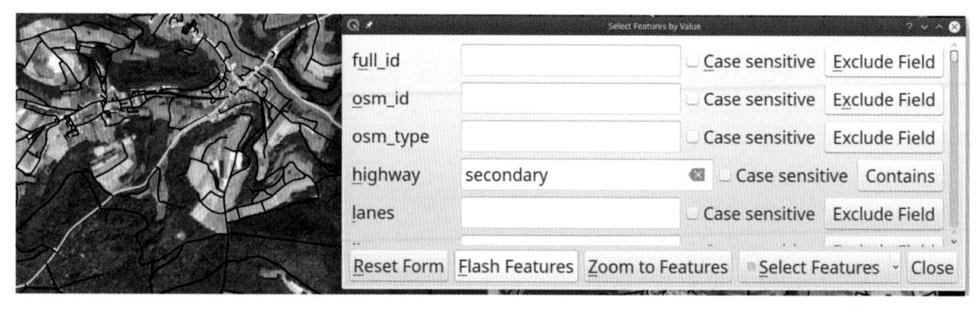

Figure 5.5 Features within the selected object can be selected using the Select feature by value interface in the Attributes toolbar. In this example, all features with the secondary attribute in the highway column are selected.

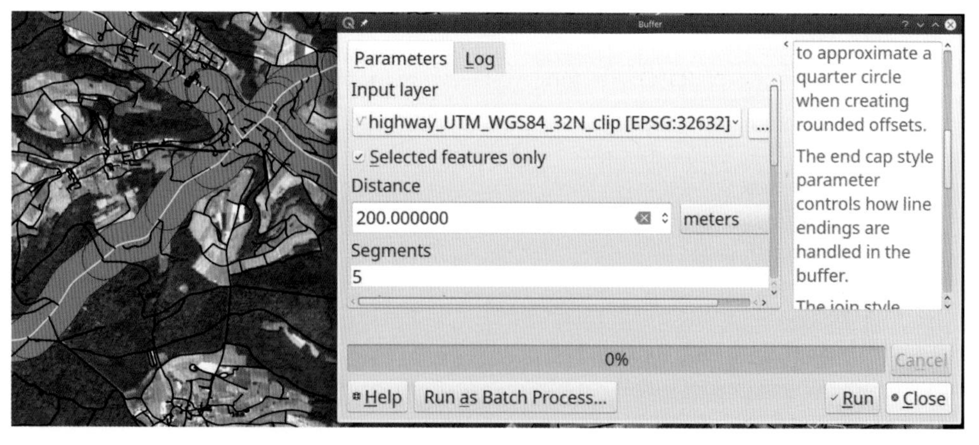

Figure 5.6 The selected roads are buffered by 200 m. Choosing the selected features only is important otherwise all roads will be buffered.

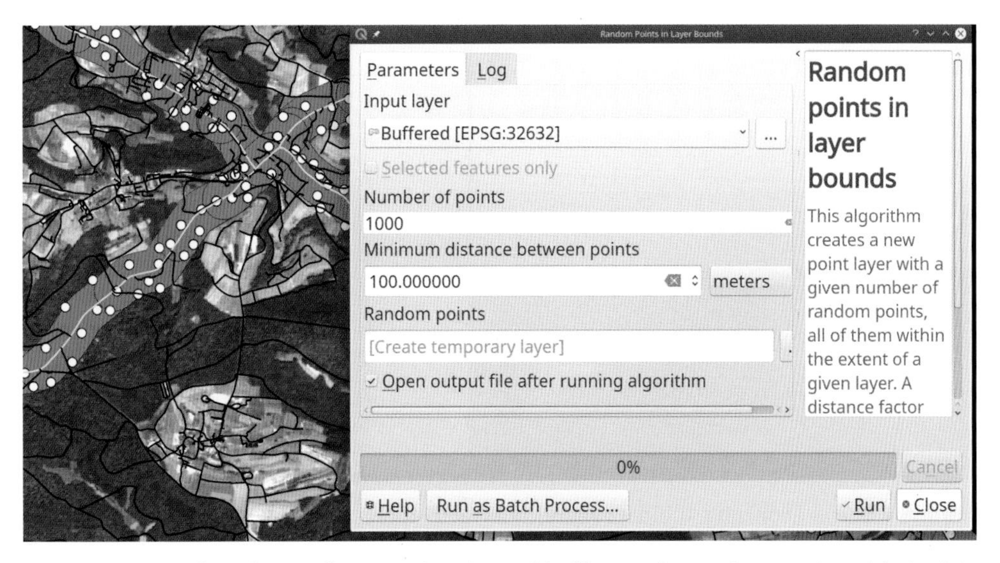

Figure 5.7 Based on the newly created main road buffer, random points can be added solely to these areas. This leads to new sampling points in the vicinity of certain road types.

constraints, extensive fieldwork far away from roads might not be feasible; therefore, you may have to select in situ sites in certain areas only.

Sampling can be constrained by creating buffers around predefined features and assigning this area to sample the in situ locations. Using the road network vector object (highway_UTM_WGS84_32N_clip) and by opening the *Select feature by value* interface via the *Attributes* toolbar, we can filter all roads that have the *secondary* attribute in the *highway* column (Figure 5.5). This selection is buffered (*Vector > Geoprocessing Tools > Buffer ...*) by 200 m. Note that we selected the *Selected features only*, otherwise all roads would be buffered (Figure 5.6). You might also want to select *dissolve*, so that no single buffers are present but overlapping buffers are combined. The resulting vector object is then used to place random points within this buffer of 200 m around the selected road types (Figure 5.7). Of course, this buffer merged with further buffers, e.g. of settlements, can be inverted and used for placing random point not close to roads and settlements.

5.2 From GIS to global positioning system (GPS)

Once the sampling points have been set, the coordinates must be exported to a GPS to navigate towards your predefined locations. This can be done by saving your points in GPX file format (Figure 5.8) and copying the file into the respective folder of your GPS device (or any handheld device). It is important to set the coordinate reference system (CRS) to geographical coordinates (EPSG: 4326) and change the *GPS_Use_Extension* to *YES*. Keeping the *Add saved file to map* option will result in an error message because GPX files cannot be added automatically. In any case, we do not want to add this new file to our project. Instead, we want to copy the new GPX file to our GPS. In our case, we must open the GPS device like any other storage volume and navigate to the GPX folder. Adding the GPX file to this folder will display the points as waypoints after disconnecting the GPS and restarting it.

An alternative option to transfer the points to our GPS is to use the *GPS Tools* within

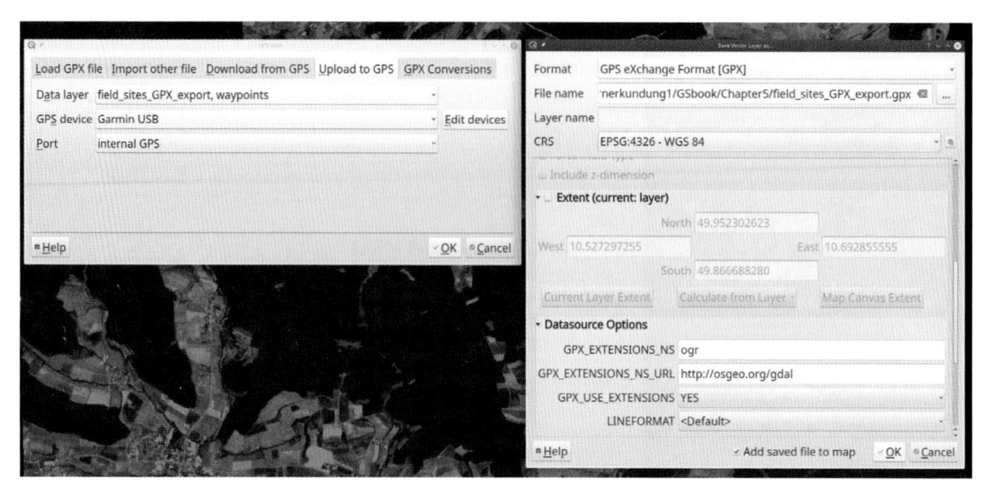

Figure 5.8 The Digitizing toolbar where, after activation, new features can be added, or selected ones can be deleted or copied.

QGIS when the GPS device is connected to your computer via a USB or serial cable. The *GPS Tools* interface allows you to convert, upload and download points and tracks to and from your GPS (Figure 5.8). The spatial vector file must be saved as a GPX file (see earlier in the chapter for the procedure required) and then loaded into *GPS Tools*. Afterwards, the selected data layers created from the imported GPX file can then be uploaded to the GPS.

In both cases, we highly recommend that you check the imported data on your GPS. Open the *Waypoint Manager* and check the coordinates and locations on the map. The worst mistakes are those you do not recognize immediately, for example, slightly shifted locations that look correct at first glance but are in fact false. Since something like this could compromise an entire field campaign, you should always critically evaluate any action affecting your data, including conversions and imports/exports.

5.3 On-screen digitization

Sometimes, all or some of the sampling data need to be collected from within a GIS. Compared to *field sampling*, this process is called *on-screen digitization*. It can be used to create *pseudo field data* for areas that are not accessible or supplement field data with additional sampling sites to enforce homogeneous sampling of your landscape. *Pseudo field data* can be created using additional information contained within very-high-resolution spatial imagery, for example, aerial footage or satellite sensor data. Besides the creation of on-screen *pseudo field data*, *on-screen digitization* can be used to map new spatial information not contained in existing data products, such as roads. Environmental data available through data portals or collaborators might not always be enough; therefore, the generation of study-specific, accurate data through *on-screen digitization* might be necessary.

In general, data obtained from portals or other researchers are preferable because they mostly provide highly accurate and sound spatial information that would be difficult for you to achieve through digitization. For example, national park boundaries are available through Protected Planet (https://www.protectedplanet.net). Such data would be difficult for you to obtain. However, in certain cases, some relevant features are inaccurate, missing or not yet available but have been mapped during fieldwork or can be seen on high-resolution imagery. For example, the land cover vector polygon extracted from OpenStreetMap (*landuse_UTM_WGS84_32N_clip.shp*) includes areas that have not been digitized by the OpenStreetMap community (Figure 5.9). Thus, such missing areas could be, if needed, added by you by activating the editing of the vector layer and adding new polygons for the missing areas. Of course, you must know the land cover to be assigned to these areas to render the modification in a useful way. Therefore, you must create your own spatial vector object that holds this type of information within QGIS:

```
Layer > Create Layer > New GeoPackage Layer ...
```

Besides naming the new file, we must define the *geometry type*, which is usually a point, line or polygon. The choice of which type depends on your aim(s). If you want to create road information, you should choose *lines*. For land cover or building extents, a *polygon* type would be more appropriate. Geometry types can then be converted to other types if needed. An important, but often neglected item, is the projection. Here, you define in which projection the new file should be created, which is normally the same projection you chose for all your other data sets. The drop-down menu lists some recently chosen projections as well as the project CRS. In the last section, new fields (or columns) of your vector can be defined. The type (text or numbers) determines what kind of data you can enter into a column. Therefore, if the column contains the name of your site, it needs to be of type *text*. If you want to add measurements, then either *whole* or *decimal* number needs to be set as the type.

For example, if you want to digitize different study areas, you should select polygons as *geometry type*, select the projection you defined for your study, for example, UTM WGS 84 and then add as many columns with specific types as you need. Usually, a column

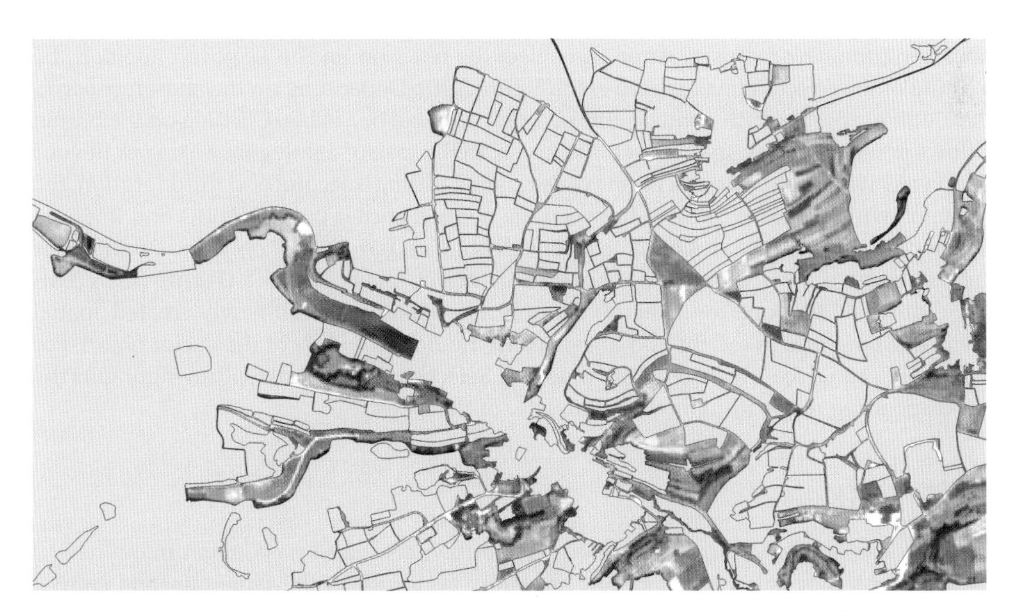

Figure 5.9 The downloaded OpenStreetMap land use vector polygon object does not cover all areas in your study area. These could be filled by you through on-screen digitization and by assigning the correct land cover in the Attribute Table.

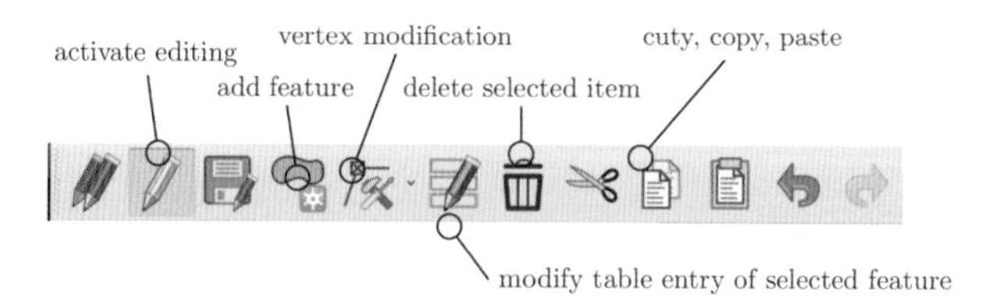

Figure 5.10 The digitizing toolbar where after activation new features can be added, or selected ones can be deleted or copied.

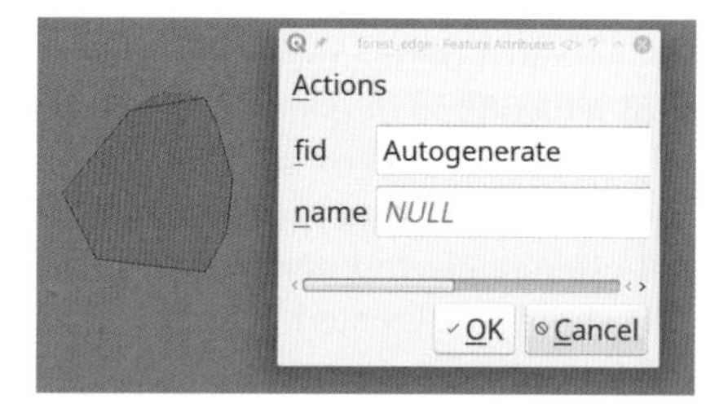

Figure 5.11 After each spatial feature is added, the attributes of each feature can be added in the pop-up window. Values can be added or changed afterwards in the Attribute Table itself.

with the names of the areas (type: text) is needed but also maybe the date when it was established and several numeric type columns that hold the existing data sets for the study area. If you wish to digitize land cover classes, the *Attribute Table* should have class name and class number columns. Columns and projection can be changed thereafter, whereas the *geometry type* cannot.

After clicking on *OK*, a new entry in your *Layers Panel* will appear. You can then start digitizing the new features. After you select the *pen* icon on the *Digitizing* toolbar, the editing of the selected layer is activated (Figure 5.10). Depending on the chosen *geometry type*, either a button with points, a line or a polygon will be visible. Clicking on the button allows you to add new spatial vector features. In the case of a polygon, a *left-click* in your map view will create a first location (vertex); after adding a second and third location, the polygon will appear as the area between the defined corner points. Points and lines are created similarly. If the desired vector object is created, *right-click* on it and an interface appears where the attributes for this specific object can be entered (Figure 5.11). This procedure can be done for as many features as needed. For each newly created spatial item, attributes are asked for and are saved in the *Attribute Table* of the vector object. For each new spatial feature, a new row is added. Selecting one spatial feature highlights the corresponding row within the *Attribute Table* of that object.

> ## TASK
>
> QGIS also allows you to select features and then copy and paste them into a new vector object or delete them. Create an empty vector file of the same type and copy some polygons from one layer to another.

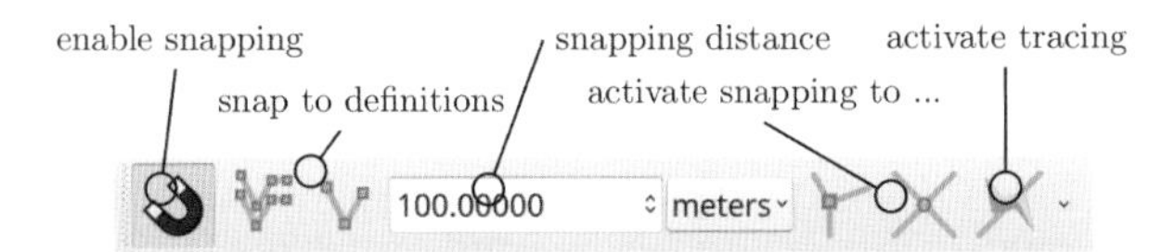

Figure 5.12 The Snapping toolbar allows you to define the distance and location to snap to. The tracing option automatically creates a vector along the shape of an existing one; thus, you do not need to follow its shape manually.

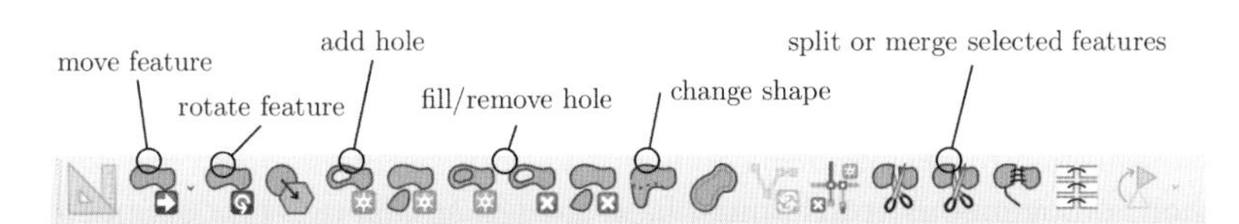

Figure 5.13 The Advanced Digitizing toolbar allows more sophisticated modifications of your vector objects, such as splitting, merging or filling holes.

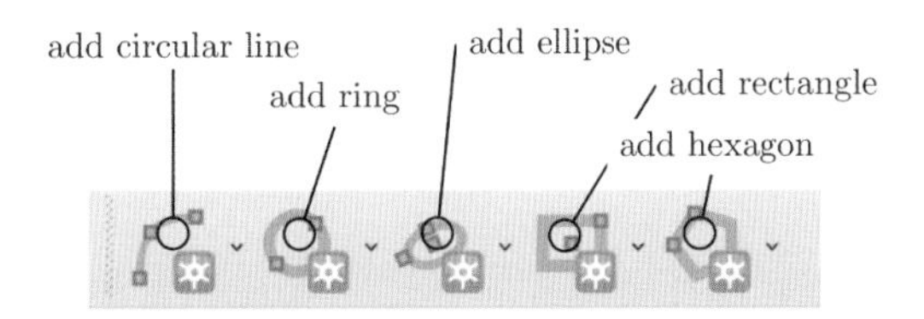

Figure 5.14 The Shape Digitizing toolbar allows you to create specific shapes, such as ellipses or circles.

Additional toolbars for more advanced operations and functions, which ease certain modifications, are the *Snapping* toolbar (Figure 5.12), the *Advanced Digitizing* toolbar (Figure 5.13) and the *Shape Digitizing* toolbar (Figure 5.14). They provide a magnitude of functionalities that can be very advantageous for different vector creation or modification operations. For example, splitting an existing polygon into two or merging two selected polygons can be done by using the split and merge features accessible through the *Advanced Digitizing* toolbar. If you need to add new features close to the existing ones and you need them to align perfectly, you can use the *snapping* options of the *Snapping* toolbar. This way, you avoid adding features accurately along the other shape but leave it to 'jumping' (or snapping) to the closest point. Specific shapes can also be added as spatial objects, such as circles, ellipses, rectangles or hexagons, through the *Shape Digitizing* toolbar. Thus, you do not need to draw a circle manually, should you need one for your project. Further fine-tuning of new vector objects can be achieved by using the *Advanced Digitizing* tool (the icon with the ruler and square) where the angle, length and distance of the lines can be accurately defined.

5.4 Summary and further reading

We have explained the basics for planning and creating relevant spatial data for fieldwork and moving these objects to a GPS device. However, many more options exist, such as *live tracking* of your GPS positions within your GIS, weighted random points or buffers for more sophisticated spatial sampling. In Chapter 6 of our book *Remote Sensing and GIS for Ecologists*, which is aimed at intermediate spatial data users, we covered field data sampling more extensively. That chapter provides more details on sampling strategies and methods using R. Moreover, we highly recommend that you test your field data collection scheme regarding the planned analysis and check for statistical challenges. For example, if you are working with remote sensing data of a 30 or 250 m pixel size, but some of your field data (e.g. clustered sampling) are just 10 or 20 m apart and your research question is related to spectral information contained in your remote sensing data, you might encounter statistical problems. Various samples within one pixel (or field or any other homogeneous entity) must be regarded as one (n=1) independent sample. Repetitive measures to derive variability are important, but in the end, you will only have one value per pixel or field for your analysis. Thus, critically evaluating your field sampling strategy from a statistical viewpoint is highly relevant. Various books address spatial statistical approaches and many books focus on field data collection only. Apart from reading such books, it is of prime importance to test your approach and gain experiences in field data collection *and* its subsequent analysis.

6. Field sampling using a global positioning system (GPS)

Finding predefined locations in the field to inspect a site and collect data usually requires a positioning system, which allows us to navigate in the field, for example, to find a study site or locate our position on a map as well as recording the coordinates for any measurement collected in the field. The most popular modern positioning system is GPS. Similar global systems are, for example, Galileo (European Union) or GLONASS (GLObal NAvigation Satellite System; Russia). GPS receivers are built into many commonly used devices, such as mobile phones, tablets or wearables. A variety of GPS devices built for specific applications are available for professional use, including lightweight tracking devices to track animals, goods or traffic, and highly accurate handheld, battery-powered devices built to be used outdoors. Compared to mobile phones or tablets, dedicated GPS devices are very useful for fieldwork due to their robustness and simplicity of use. In the following sections, we explain the recommended tasks needed to collect field data using a handheld GPS device, although the general idea applies to any kind of GPS-capable system.

GPS is a satellite-based system that allows anyone with a GPS receiver to find their exact position anywhere in the world. GPS satellites orbit the Earth as a constellation at an altitude of approximately 20,200 km. Each satellite is continuously transmitting a radio signal containing the satellite's orbital parameters (i.e. its orbital position) and time, measured precisely using an onboard atomic clock. When turned on, the GPS receiver is listening for these GPS satellite signals. Since radio waves propagate at the constant speed of light, irrespective of the speed of the satellite itself, the time delay between the emission of the signal by the satellite's antenna and its reception by the receiver is linearly proportional to the distance between receiver and satellite. Thus, if an exact reference time is known, the GPS receiver needs to pick up a minimum of three GPS signals to precisely calculate its location on the Earth's surface, which is represented by latitude, longitude and altitude. Since handheld GPS receivers do not carry an atomic clock and have no precise time information, they require the reception of a fourth GPS satellite signal to calculate a reference time. Thus, a handheld GPS device displays and records its position in the field when a minimum of four GPS signals are received, meaning that the receiver has direct visibility to a minimum of four satellites overpassing its position. The more satellites are in sight, the more precisely we can calculate our current position.

6.1 GPS in the field

Common handheld GPS devices for fieldwork can store locations (*waypoints*), *routes* (a sequence of locations that make up a planned route) and *tracks*, which represent the receiver's movement over time. Waypoints, routes and tracks are the three basic feature types in GPS data. QGIS displays waypoints in point layers, while routes and tracks are displayed in *linestring layers*. The file format used to exchange such information is GPX.

Coordinates collected during a field data collection campaign are usually retrieved with a GPS device and stored in the device or recorded on a sheet. A certain level of redundancy is usually recommended; therefore, we advise you also always manually record the coordinates. Especially if you use a data collection sheet as described in Chapter 5, further details must be noted; thus, recording the coordinates on the same sheet as the actual measurements ensures that both pieces of information match. It is not important how the coordinates and data are collected; they can be imported into QGIS later. A text file with the coordinates or a direct link to your GPS within QGIS allows you to work with your data (Figure 6.1).

Noting down coordinates and measurements on a piece of paper and inputting the information later into a spreadsheet requires a sound understanding of some GPS issues. Generally, we highly recommend reading the manual of your device to make full use of the power of the device. Some key aspects using a commonly available GPS device are elaborated further on in the chapter.

A GPS device can be customized in various ways and each brand or version is different. Some key features that need to be checked when working with a GPS device in the field are satellite reception, your position or adding and finding a waypoint (Figure 6.2). Satellite reception impacts the availability of (accurate) coordinate information and differs depending on your location. Dense forests with partly closed canopies might hamper the signal while open landscapes lead to a better signal.

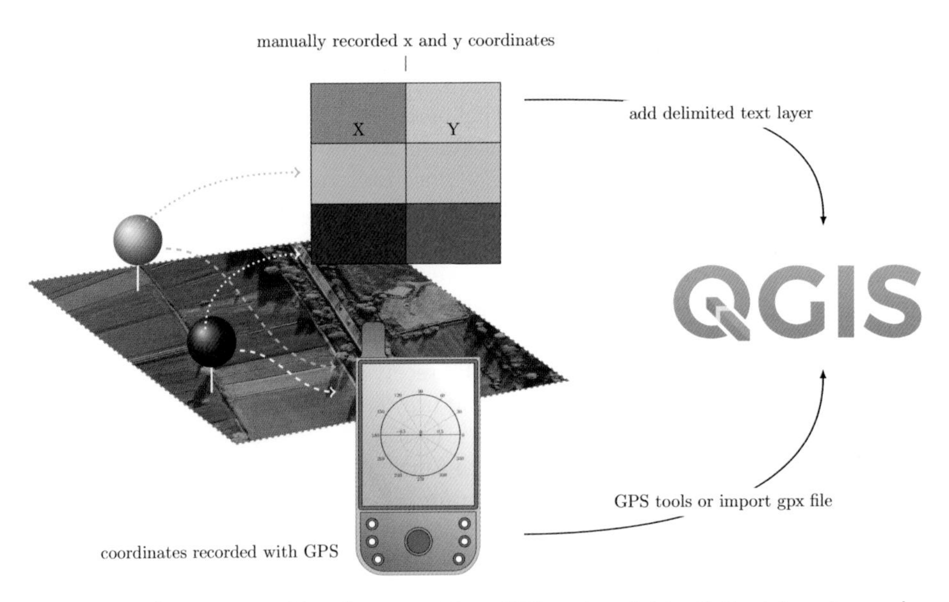

Figure 6.1 Coordinates can either be stored in a GPS or in a table. Both data sets can be imported into a GIS.

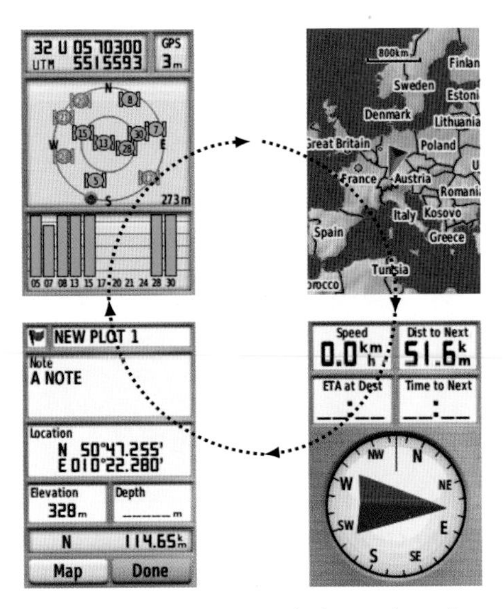

Figure 6.2 Various displays or functions are regularly used, such as map viewing, adding or finding waypoints and checking satellite reception.

For this book, we have used a GARMIN GPS device. Other models and brands have similar interfaces. After turning the GPS device on, you are presented with various options to navigate through several screens using keys such as *page*. (Other brands might have a different user interface, but base functionalities are the same.) After navigating to the *Main Menu* and selecting it using *Enter*, you are presented with several options to customize your GPS settings. Selecting *setup* leads to another set of options. Here, the *position format* (coordinate reference system (CRS)) can be selected (Figure 6.3). The projection interface is very important because it defines the CRS where collected locations

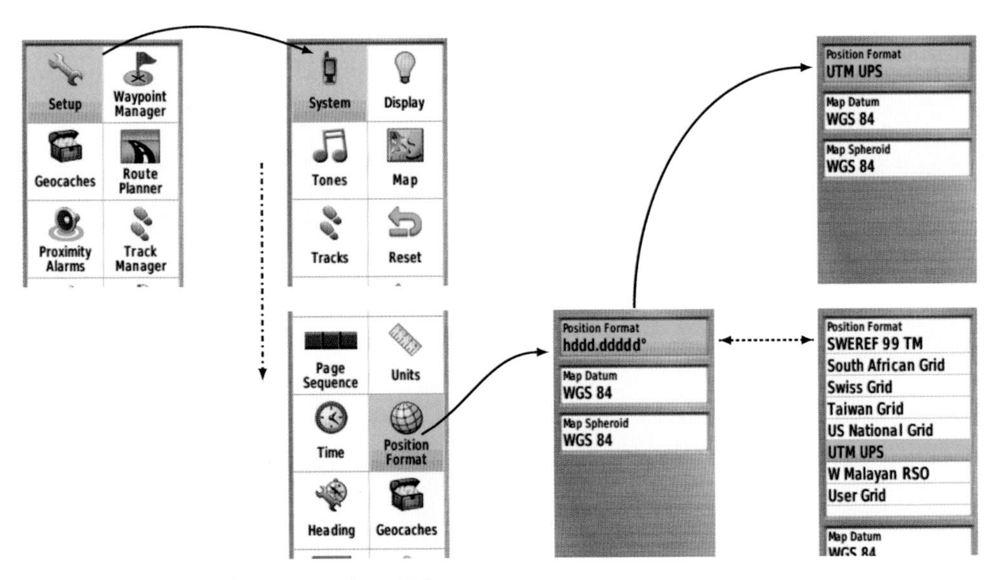

Figure 6.3 Setting the CRS inside a GPS.

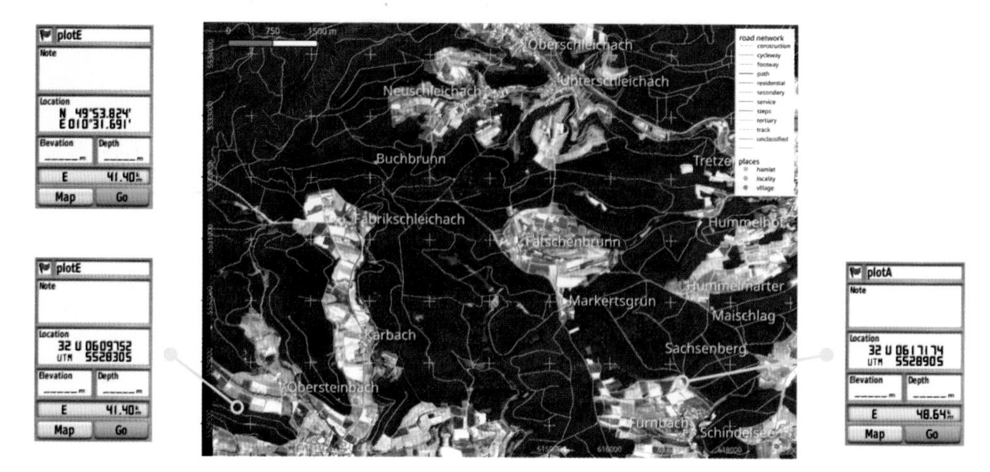

Figure 6.4 The fieldwork map created previously with the coordinates shown in the GPS device. Note that changing the CRS to the same projection used in our QGIS project is relevant to match the coordinates shown in the GPS device and on the map.

should be recorded. The default is a geographic coordinate system that we change to the CRS *UTM WGS84*, as shown in Figure 6.3, by selecting the right entry with the *up* and *down* keys, and confirming the selection using the *Enter* key on the GPS device. One reason for changing the CRS to the same one used in our QGIS project is the orientation on our field map. Simply put, using geographical coordinates would not allow us to find our position on the map. Changing it to the same projection allows us to match the coordinates depicted in the GPS device and on the map (Figure 6.4).

Finding field sites is a key task in any fieldwork. Selecting the waypoints of your locations and navigating via a map or compass to that point is usually done (Figure 6.5).

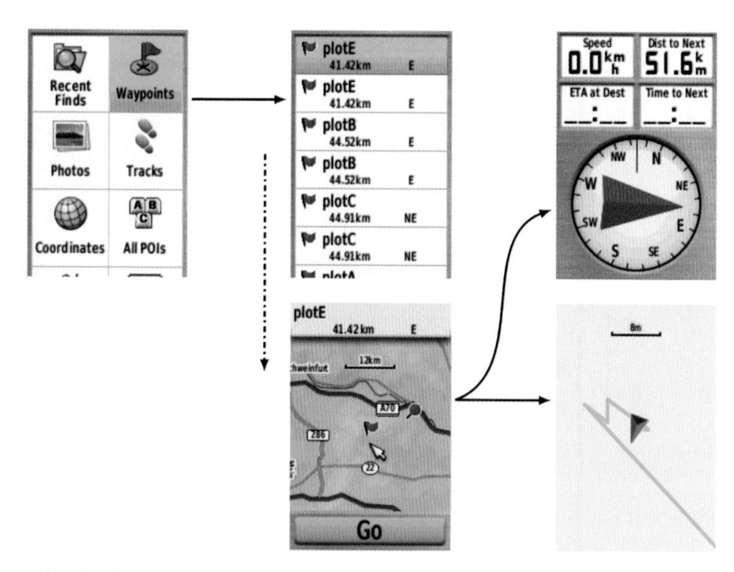

Figure 6.5 Individual waypoints can be selected. The GPS device will help you to navigate towards them using a compass or a map.

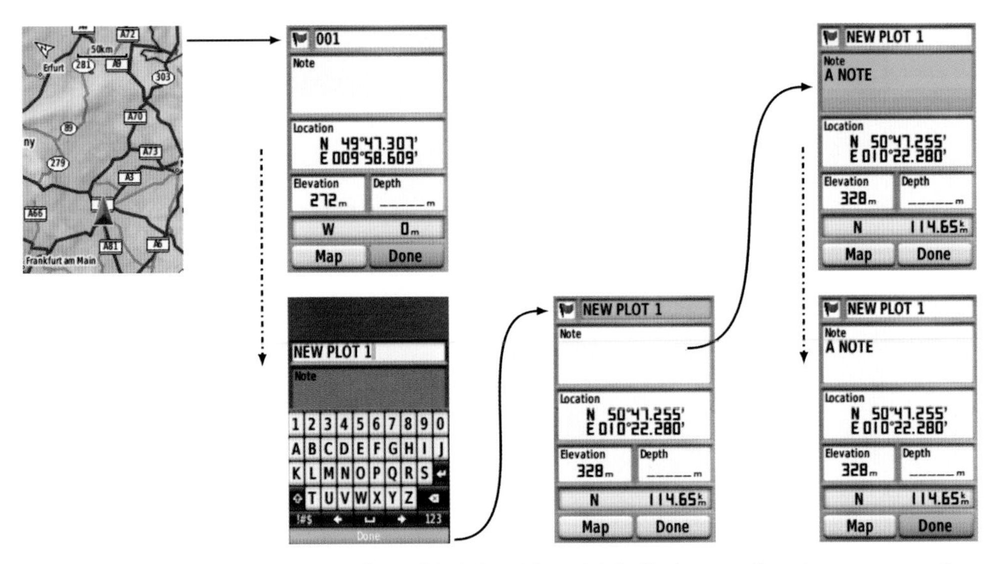

Figure 6.6 New waypoints can be added (mark) and labelled according to your sampling scheme; notes can also be added.

Setting a new waypoint and adding a name and further remarks apart from the coordinates is a common task. This can be done with various devices; how the information is added depends on the device used. In our case, a virtual keyboard is shown, which is used to select single characters to write the site name (Figure 6.6). Besides setting waypoints for new sampling locations, it is also highly advisable to set the waypoints of your car position or research station. In some landscapes, it is very challenging to find your car, tent or research station, especially if there is no cellular connection to use an online map service. Therefore, always mark the position before you leave. Sometimes, some intermediate landmarks should also be marked for the purpose of better orientation.

6.2 GPX from GPS

After the field campaign, data are usually entered into a spreadsheet or exported from your handheld device. If the coordinates are recorded in a spreadsheet, the coordinate columns can be used to import the data into a geographic information system (GIS) and convert them to a spatial object. Remember to note the projection in your spreadsheet, otherwise the imported coordinates cannot be assigned to the correct CRS. Alternatively, the GPX file can either be retrieved through the *GPS Tools* within QGIS or the GPS device can be accessed like any other storage device and the GPX file stored in the GPX folder can be copied and imported into QGIS.

Nowadays, a classic GPS device is not always necessary due to the availability of GPS receivers in mobile phones and tablets. Various software programs built for portable devices are on offer. Searching the app store of your mobile phone for a mobile GIS might lead to a useful and pragmatic solution for your fieldwork. An overview can be found on the OSGeo GIS Mobile Comparison wiki (https://wiki.osgeo.org/wiki/GIS_Mobile_Comparison); several experimental and stable QGIS versions for Android are also available (Figure 6.7).

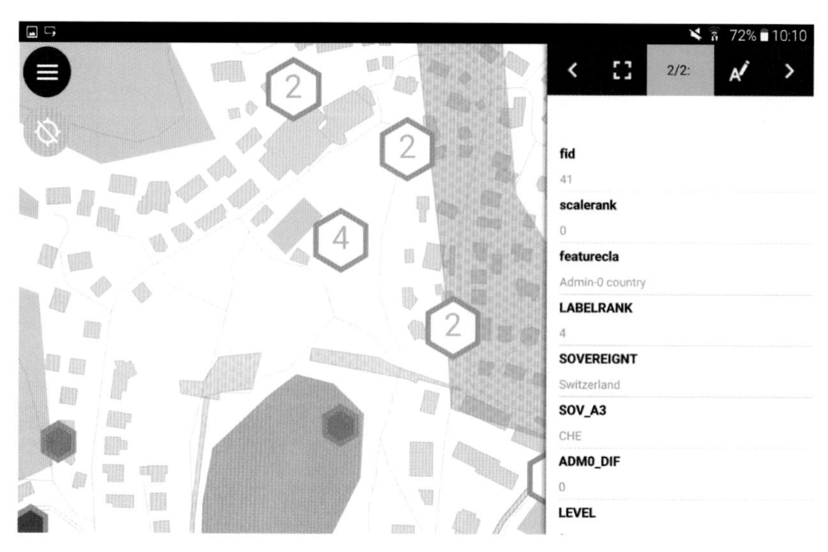

Figure 6.7 The QField mobile QGIS app allows you to navigate on a map and enter new field data. Apart from known table attributes, pictures can also be added.

6.3 Summary

Space-borne positioning systems are an important feature of in situ data acquisition for the environmental sciences. Without free and openly available GPS radio signals that can be received around the globe, essential ground truth information needed for various applications would be missing. However, being able to locate yourself in the field is not the only thing needed to successfully collect field data: a software interface that allows you to set waypoints and plan walkways efficiently is helpful when conducting fieldwork. Dedicated GPS devices are professional tools for such a task. In mobile phones, positioning technology is combined with internet connectivity, offering the possibility to use mobile GIS applications in the field.

7. From global positioning system (GPS) to geographic information system (GIS)

The final part of any fieldwork is to convert field samples into a digital format and into the spatial domain, namely a GIS. Usually, data are collected on sheets and the corresponding site name or coordinates are recorded. If data are collected at a predefined site, either the coordinates or the spatial vector file already exist; thus, only the measurements at each site and the site names must be linked. If the site coordinates are recorded on the same sheet as the measurements, this table can be imported and converted to a spatial object using the column that holds the coordinate values (Figure 7.1). However, for this approach to work, it is crucial to know the coordinate system used. Otherwise, the coordinate information is of no value. The last option is to download the coordinates from a GPS device and link them to the corresponding ID in the measurement data set.

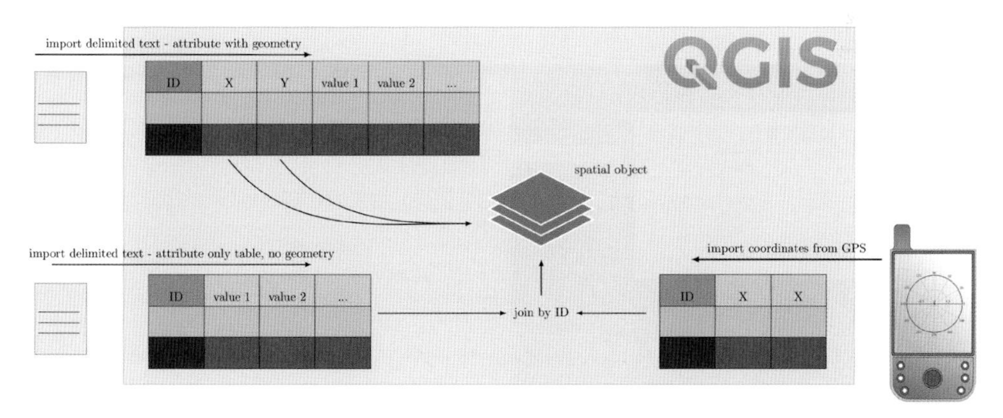

Figure 7.1 The coordinates need to be matched with the corresponding measurements which are either already listed within one table or need to be joined by a common ID within QGIS.

7.1 Joint coordinates and measurement sheet

In most cases, a spreadsheet is created where the coordinates, site name and corresponding measurements are recorded. This file can be imported as a comma-separated value (CSV) file into QGIS using the *Add Delimited Text Layer* button (which looks like a comma) or via:

```
Layer > Add Layer > Add Delimited Text Layer ...
```

We import the coordinates and corresponding attributes of the file *coordinates_with_attributes.csv* as *point coordinates* for geometry definition and most important for projection (*Geometry CRS*). We know that we recorded the coordinates of our field sites as *UTM WGS84 Zone 32N*; thus, we must choose this coordinate reference system (CRS) in the *Geometry CRS* field. The corresponding column names must be selected for the x and y coordinates (Figure 7.2). After adding this text file, a new layer is added to your list of layers, whose styling you may change, as already described in Chapter 1. The layer is still only stored as a text file that is spatially interpreted by QGIS but is not physically present as a spatial object. Therefore, we need to export it as spatial object, for example, as a GeoPackage file, to our file system for later usage:

```
Right-click on layer > Export > Save Feature As ...
```

Within the export interface, you can define which columns should be saved if you do not want to keep the coordinate columns.

> ### TASK
>
> Check what happens if you swap the x and y coordinate column in the delimited text import interface. What happens if you choose the wrong Geometry CRS?

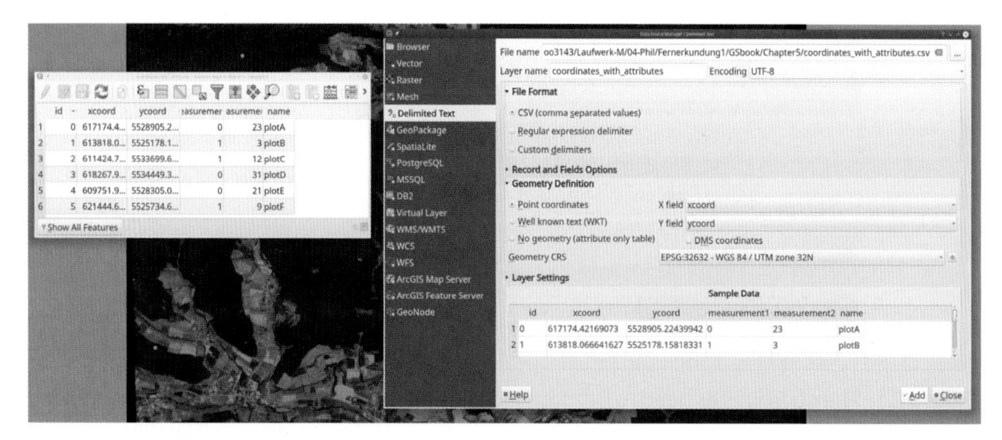

Figure 7.2 The delimited text layer import interface and resulting spatial points, and its corresponding table, are shown.

7.2 Separate coordinates and measurement sheet

A slightly more complex procedure must be followed if your coordinates and measurements are available as different files. Coordinates can be stored in a spreadsheet; thus, we can import them either as a delimited text layer, as described earlier, or as a spatial object, either because we created the (random) points ourselves beforehand or we already imported a GPS file. In both cases, coordinates must be joined with the actual measurements to bring our location attributes into the spatial domain. This also requires that, in both objects, a column with an identifier is present, such as the site name or a location ID that links coordinates and attributes.

First, we import the file containing the coordinates only ('coordinates_only.csv') via the delimited text layer import function as explained earlier. Then, the same import interface is used to import the attributes file ('attributes_only.csv'), the only difference being that we select 'no geometry (attribute table only)' within the *Geometry definition* section. The two imported layers are listed in the *Layers Panel*. The different icon identifies the *attribute_only* layer as *non-spatial*. Within the *Properties* dialogue window of the *coordinates_only* layer (double-click on the layer or right-click > *Properties*), we select the tab *Joins* (Figure 7.3) and add a new object to be joined by selecting the *plus* sign. The *Join Field* and *Target Field* are essential for this operation. Using these fields, you can define the names of those columns in your tables that should be linked. The two column names do not have to be equal; the referenced columns should simply represent corresponding values or characters, for example, an ID, which identifies matching entries in both tables, so that the coordinate and attribute files can be joined correctly. Within this interface, you can also define which fields should be joined. After selecting *OK* and opening the table of the *coordinates_only* layer, the attributes of the *attributes_only* file are listed; thus, they are now spatially located (Figure 7.4). So far, the *coordinates_only* file is still just a CSV file

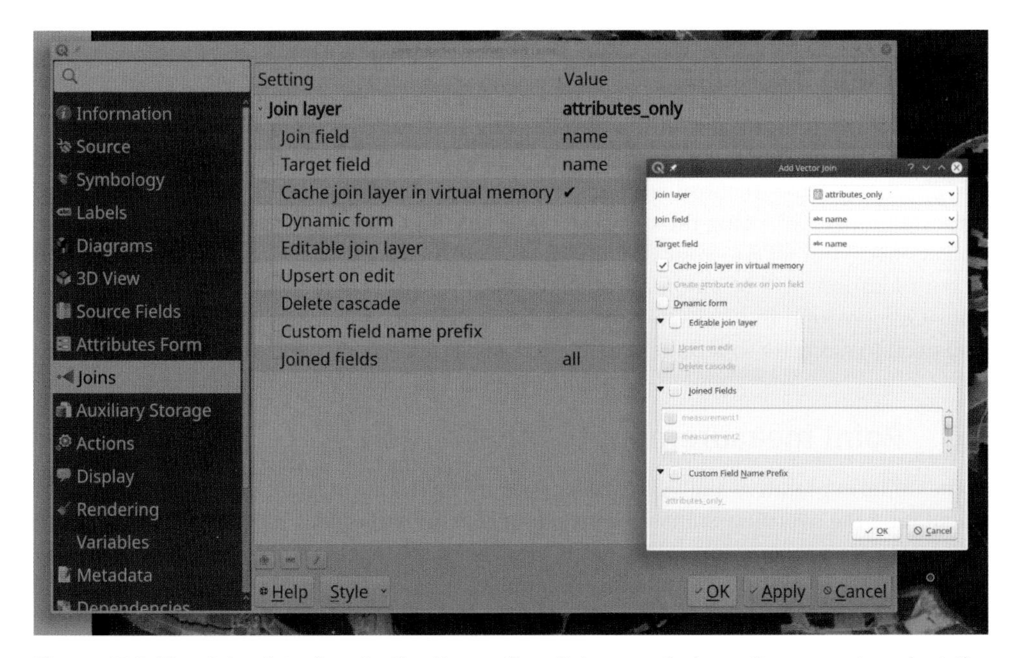

Figure 7.3 The Joins interface in the Properties dialogue window allows you to select the table to be joined and requires matching identifiers in the columns of both objects.

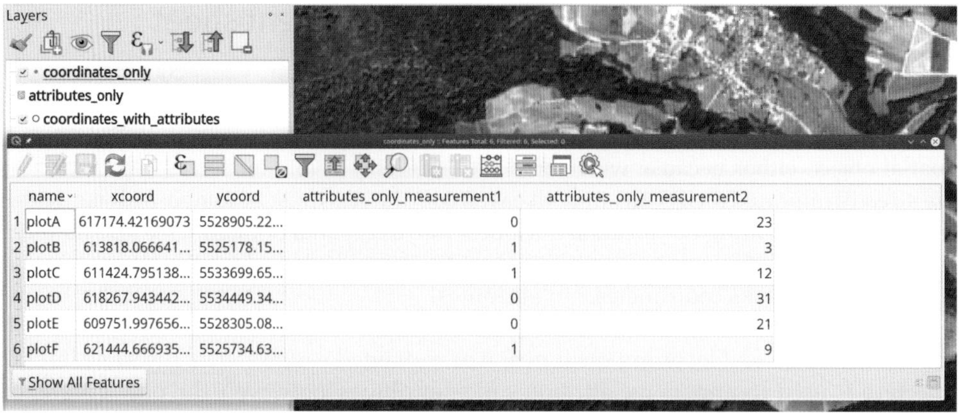

Layers
coordinates_only
attributes_only
coordinates_with_attributes

	name	xcoord	ycoord	attributes_only_measurement1	attributes_only_measurement2
1	plotA	617174.42169073	5528905.22...	0	23
2	plotB	613818.066641...	5525178.15...	1	3
3	plotC	611424.795138...	5533699.65...	1	12
4	plotD	618267.943442...	5534449.34...	0	31
5	plotE	609751.997656...	5528305.08...	0	21
6	plotF	621444.666935...	5525734.63...	1	9

Show All Features

Figure 7.4 The resulting table of the Joins attribute is now associated with the spatial object.

and would lose its spatial properties once QGIS is closed. The layer must now be saved as a spatial object to store the coordinates and attributes together (*Export > Save Features As ...*).

Once the in situ measurements are linked with a corresponding spatial object and saved as a spatial object, we can modify them if needed. The recently exported file now stores a normal vector object and can thus be modified. Single points can be selected, their attributes can be changed, or their spatial location can be adjusted using the *Move Feature* option within the *Advanced Digitizing* toolbar. We highly recommend to critically evaluate the location of the in situ data (in case the coordinates were imported) and the corresponding values joined with the spatial object to minimize the risk that false coordinates are imported or that table joins falsify any subsequent analysis.

7.2.1 Further options

The *GPS Tools* plug-in or additional GPS plug-ins, such as *Batch GPS Importer*, can be imported into QGIS. If you do not want to link your GPS directly with your GIS, you can also browse the memory of your GPS and copy the GPX file. This file format can be imported into QGIS using the *GPS Tools* interface and then saved as GeoPackage. When you connect your GPS to your GIS and download the coordinates directly, the projection of your locations will be automatically converted to your project CRS. You can also track your GPS position in your QGIS project using GPS tracking. This might be relevant if you need to navigate to certain landscape features and check your map simultaneously.

7.3 Point measurement to information

Some initial explorative analysis can be executed on the imported measurements, such as calculating the overview statistics for each field. This serves two purposes: first, to check if the data are properly imported and if they make sense; second, to get a first overview of the properties of the field data measurements.

Some simple statistics can be retrieved executing the *Basic Statistics for fields* operation:

```
Vector > Analysis Tools > Basic Statistics for Fields ...
```

Using this operation, you can define the field needed to calculate the statistics. The output is an HTML file with information such as *count of unique values* or *most frequently occurring value* of the values within the selected column.

A distance matrix of your in situ data points and other point layers can also be computed using:

```
Vector > Analysis Tools > Distance Matrix ...
```

The output gives the distance between each of your field data locations and each point in the second layer (the second layer could be settlements). The matrix provides a distance matrix between all your locations and all settlements. Further details on vector analysis are addressed in Chapter 8.

The data sampled in situ can also be displayed on the map or using the *Data Plotly* plug-in introduced in Chapter 4. The *Data Plotly* functions allow you to create informative graphs of the used data and link such graphs with the corresponding spatial objects. Selecting points within a graph highlights the corresponding locations on the map. Moreover, a simple display of values can be achieved by plotting different sizes of circles based on the column values rather than colours, as shown in Figure 7.5. The graduated styling is changed from *colour* to *size* in the method section. The minimum and maximum size of the circles can also be defined, as well as the colours of the number of classes.

Additionally, it is possible to check the distribution of the field samples compared to the land cover information (landuse_UTM_WGS84_32N_clip.shp). This gives us an idea whether all land cover polygons have been equally sampled, which is unlikely with a sample size of five. Of course, we could have distributed our random points in a more stratified fashion based on land cover (as described in Chapters 5 and 6). However, let us assume we have sampled the locations in the field and would like to check how well they represent our land cover polygon distribution. We apply a simple check by counting the number of points per polygon:

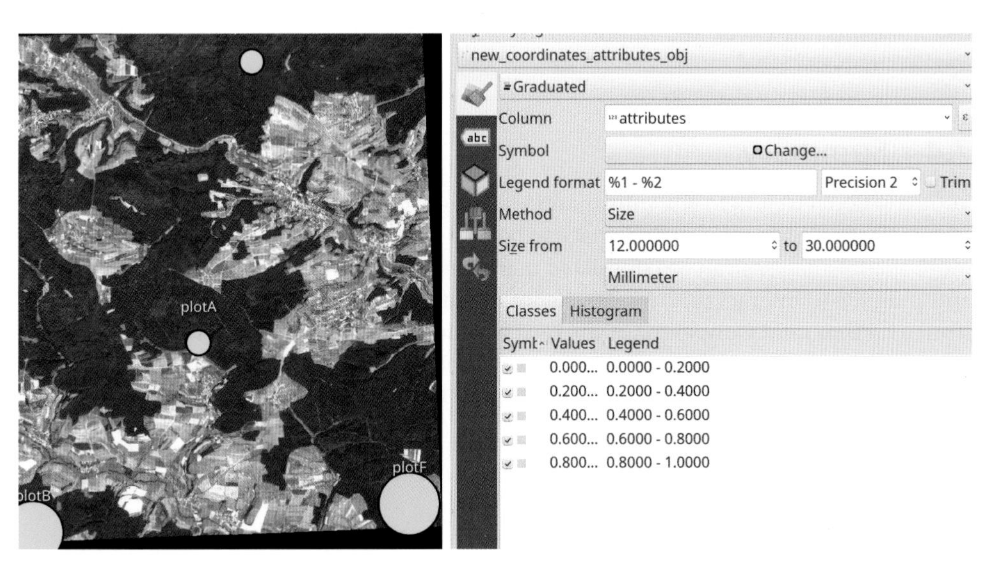

Figure 7.5 The attributes measured for each location can also be displayed in different sizes. This allows you to get a good visual overview of value distribution in the study area.

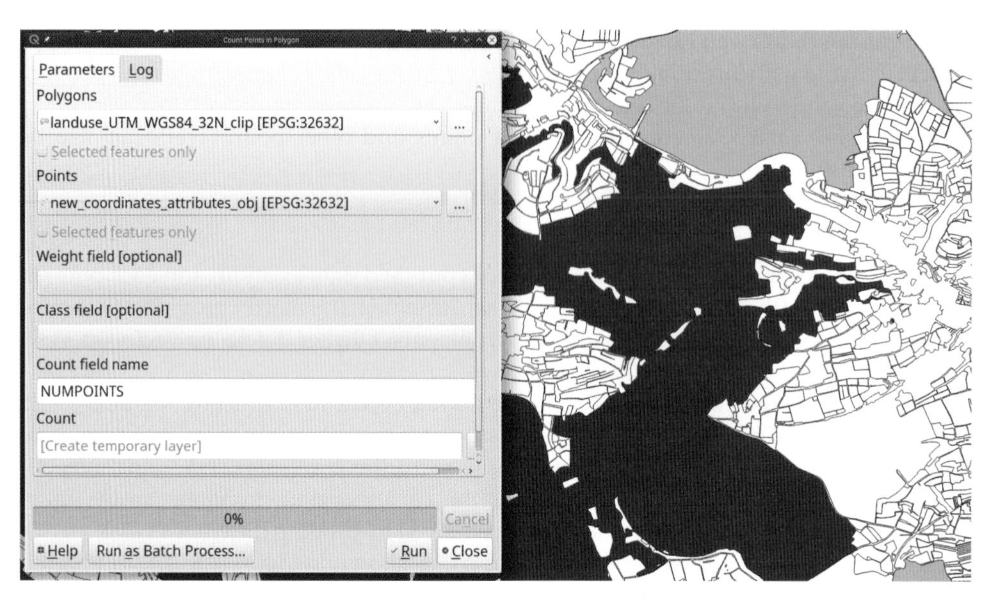

Figure 7.6 The interface used to count the number of points within polygons and the resulting vector object are shown. It is evident that the field samples have not been sampled homogeneously across all land use type polygons. A few polygons hold most of the points.

```
Vector > Analysis Tools > Count Points in Polygon ...
```

The resulting map shows that we have a strong bias and did not equally sample all landscape features (Figure 7.6). By chance, the largest polygon also had a higher count. Correcting for such differences in the landscape (e.g. class sizes) must be done beforehand or identified and dealt with afterwards, as shown in this example.

The import and evaluation of field data is important to ensure sound data for further analysis. An initial explorative analysis and display of the in situ data is always recommended to check for import or data collection issues before proceeding with further analysis.

7.4 Summary

You have converted field site information into a spatial object and learnt that they are in fact normal spatial vector objects that can be analysed like any other spatial object. Many more spatial analyses can be conducted using them. Also, you should try other software and plug-ins to interact with a GPS device and analyse or preprocess the data obtained from a GPS or field campaign. In the next chapters, you can continue to use these data sets for the analysis, such as creating a buffer around vector objects or intersecting them, as well as querying environmental values inside these field sites.

Part III

Data analysis and new spatial information

In Part III, we cover data analysis, including vector data modification and raster data analysis. We create spectral indices based on remote sensing data and intersect them with vector data to retrieve relevant environmental information for a specific research question.

This part of the book covers the fundamental steps needed in many research projects that use spatial data. Within the course of a research project, more sophisticated approaches may be required but these can be achieved by building onto the fundamental approaches provided in the book.

8. Vector data analysis

Common analyses of vector data are operations such as buffering and intersecting to gain further insights into spatial data attributes. However, in most cases more complex spatial operations are required and are very specific to the individual research project. The origin of the vector data, such as in situ data sampling, on-screen digitized vector objects or downloaded data are irrelevant to the analysis. In the following sections, we highlight several functions that do not cover the full potential of the vector modification options but provide a good overview and starting point to analyse vector data.

Extracting basic statistical data or general information can be achieved using a variety of operations. For example, retrieving basic statistics for your attribute table or knowing what the unique values are is feasible through:

```
Vector > Analysis Tools > Basic Statistics for Fields ...
Vector > Analysis Tools > List Unique Values ...
```

However, both operations do not result in a spatial output but a table or report file. Adding geometry information, such as area or perimeter, can be achieved with:

```
Vector > Geometry Tools > Add Geometry Attributes ...
```

To extract line length, the following function should be executed:

```
Vector > Analysis Tools > Sum Line Lengths ...
```

Other information, such as line crossings or intersections, can be retrieved. Using two different vector line objects, such as movement tracks and roads, we get line crossings; if the same vector object is used twice, we get actual intersections, for example, for roads. The *Line Intersections* operation asks you to input and intersect layers:

```
Vector > Analysis Tools > Line Intersections ...
```

In our example, we use the same data twice and get all road intersections in the 'highway_UTM_WGS84N_clip' data set (Figure 8.1). The resulting object is a point vector layer indicating all line crossings in this data set.

Field sample locations are often 'buffered', which means that an area around each location is created. This is done, for example, to cover spatial variability around a location and intersect the resulting buffer with land cover information. The size of a buffer depends on your research question and what you aim to achieve with such an analysis; buffer sizes can range from metres to kilometres. A static buffer size, as introduced in Chapter 1, can be achieved by executing the *Buffer* operation within the *Geoprocessing Tools* of the

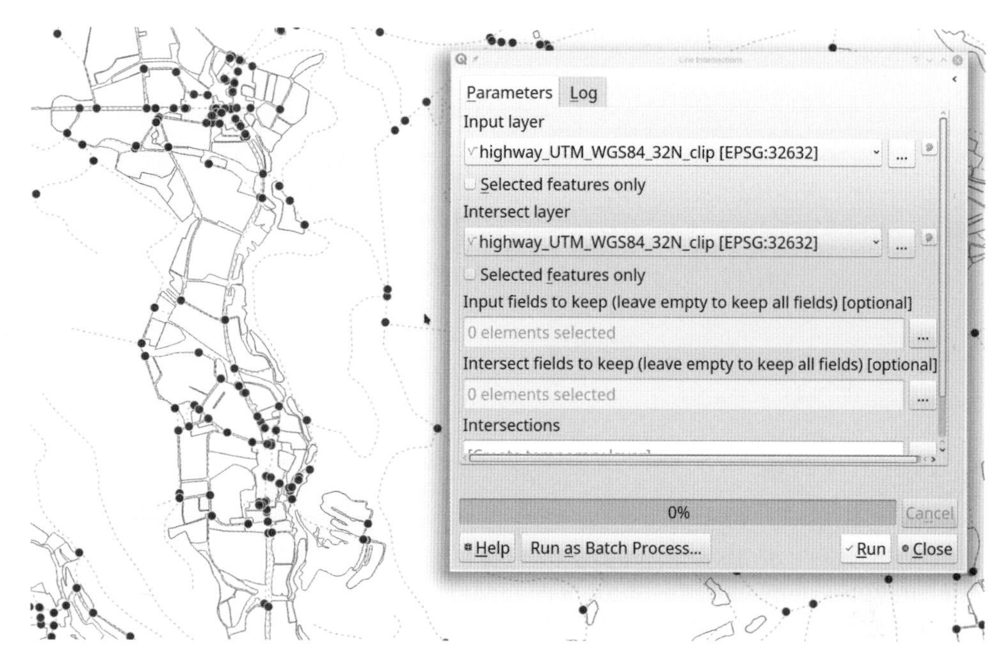

Figure 8.1 Line intersections for the same vector line object. Road intersections are displayed as points. In the background, the road input layer is displayed as a dashed line.

vector menu. However, for certain analyses a variable buffer relating to a column value in the attribute table is needed. Such functionality is provided by the *MMQGIS* plug-in. After the plug-in has been installed, a range of new functionalities are available, such as an alternative buffer function that allows you to use attributes to define the radius of each buffer. This function uses the values of one column to set the size of the buffer. Selecting this function offers a similar buffer operation than the default one:

```
MMQGIS > Create > Create Buffers
```

The *MMQGIS* buffer function allows you to set the radius attribute to be taken from a column within the attribute table. The values within that column can be in metres or kilometres; based on these values, a specific buffer radius is created. If we want to use the values of the *attribute_1* column, the values range between 3 and 31; thus, we will have buffers of 3–31 m or km. The first value might be too small, the latter too large for your analysis; therefore, we could create attribute values that correspond to adequate buffer radius values. By opening the attribute table of *new_coordinates_attributes_obj*, we can run the Field Calculator (the *slide rule* icon within the attribute table), which allows us to create new columns within our attribute table by executing a variety of mathematical functions. We create a new column called *bufferVal* that has the same value as *attribute_1* but multiplied by 50; thus, in the expression field we enter:

```
"attribut_1" *50
```

Searching functions or variables can also be done with the Field Calculator, which also provides a preview of the output or tells us if the expression is invalid (Figure 8.2). The new column now holds new values between 150 and 1,550, which are based on the values of the other column. In this example, these values are appropriate for buffer radii given

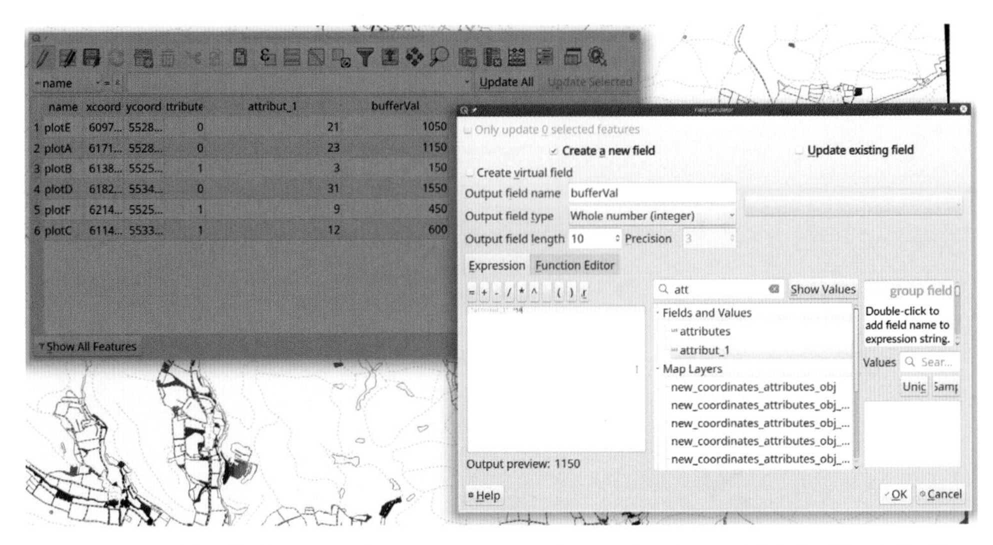

Figure 8.2 Field Calculator used to create new columns for the variable buffer definition. Attribute values are multiplied by 50 to obtain appropriate values for buffer radii given in metres. The expression used is shown and the resulting new column is shown in the background.

in metres. The Field Calculator automatically activates the editing to be able to modify the attribute table. After the Field Calculator has created the new column, the editing can be turned off again and the attribute table can be closed. Opening the *MMQGIS* buffer functions, we can now select the new attribute column and select metres as the unit (Figure 8.3). The resulting vector object shows our six field locations with variably sized buffers based on the values we created previously. As already stated, the value of the buffer sizes clearly depends on your study aims and must be defined and discussed carefully.

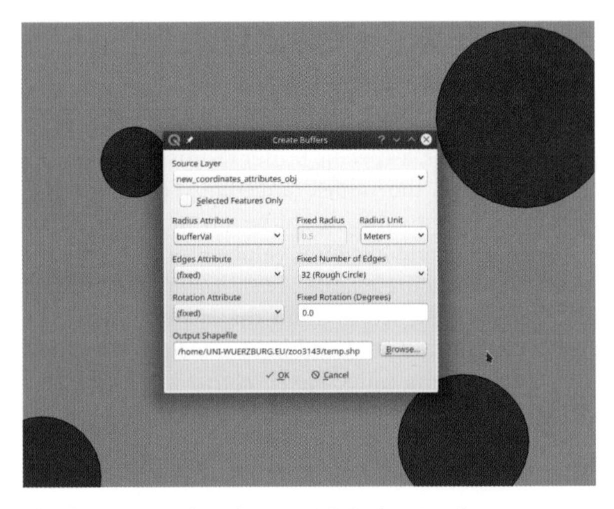

Figure 8.3 The buffer function within the MMQGIS plug-in allows us to use column values in the attribute table to define the size of each individual buffer radius. Based on the new column values created within the Field Calculator, we obtain new buffer values between 150 and 1,550, which can be used as distances in metres around our field measurements.

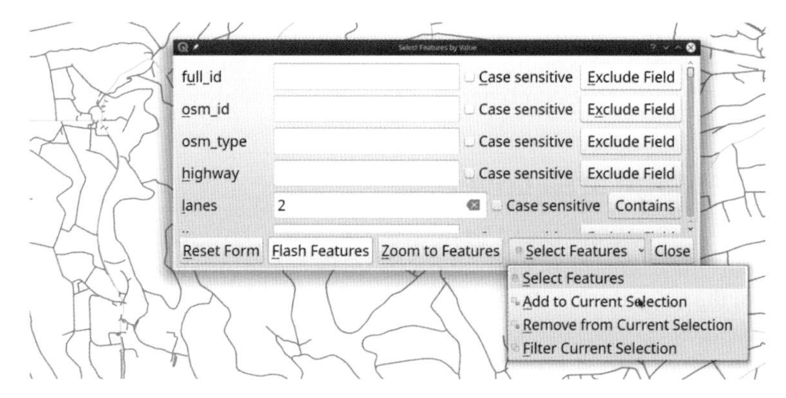

Figure 8.4 The Select Features by Attributes option allows you to select features within your vector objects that match certain attribute values. The selected features can then be used for subsequent analysis.

Another type of buffer is a multibuffer, which is a series of buffer rings for the same input data set. This is the *multi buffer* function provided by the *Multi Ring Buffer* plug-in:

```
Vector > Multi Ring Buffer
```

We execute a 500-m multiring buffer analysis on the highway data set but only for line features with 2 lanes. Therefore, we must first query the *highway_UTM_WGS84_32N_clip* vector line object for all lines with the attribute *2 lanes*. This spatial query based on an attribute can be done via the *Select Features By Value* button within the *Attributes* toolbar or by hitting F3. All lines matching the expression 'lane contains 2' are selected (Figure 8.4) and can be used for the subsequent multibuffer analysis. It is important to select *buffer only selected features* in the multibuffer window, otherwise all lines within the road vector will be buffered. The rings should have a total radius of 500 m but divided into 5 rings (Figure 8.5).

The vector operations outlined in this section provide an idea as to what is feasible

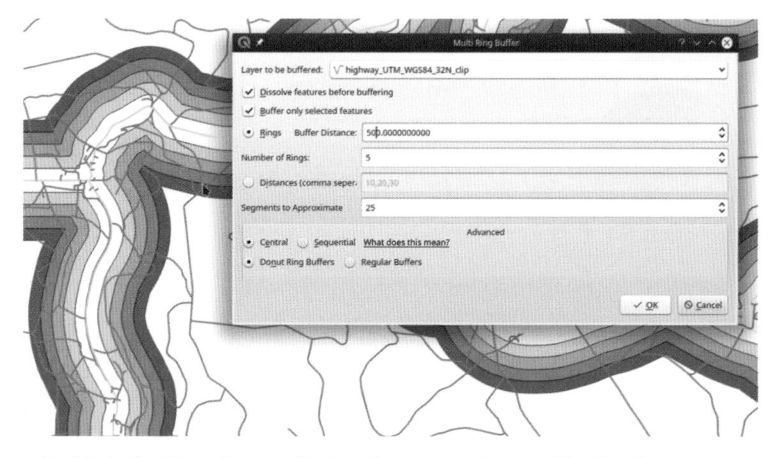

Figure 8.5 Multiple buffers for roads that have two lanes. The buffers are created using the multibuffer function in the MMQGIS plug-in. The resulting multibuffer is shown in the background in grey gradients.

using vector data. The results of such operations can be further combined and analysed to gain relevant spatial information. In the following sections, the variable point buffer and the road multibuffer are combined and their ratio is analysed.

8.1 Percentage area covered

For the analysis, we are interested in how much area is covered by the two-lane road buffers within our different point buffer sizes. We want to know the single and total percentage of the single buffer distances within the location buffers. This presents us with various challenges:

- Clipping the location buffer to the actual extent of the study area.
- Clipping the multibuffer to the buffers and keeping the individual buffers.
- The point and road buffer need to be combined to compute their ratio.
- Geometry attributes, such as area per road buffer section, need to be computed.
- The different area calculations need to be used within a calculator to calculate the percentage area values.

We start by clipping the location buffers to the actual study area extent to align it with our road network, which has been cropped to the study area. Otherwise, some road buffers might not have been created and we might include a degree of bias when computing the area of roads within the point buffers. Conversely, we could recompute the road buffers also for areas outside our study area. Both options are justifiable, but for the purpose of showing certain vector modifications, we stick with clipping the point buffers to the study area. We clip location buffers of varying sizes to the land use vector object and create semicircles for all point buffer locations extending beyond the study area (Figure 8.6):

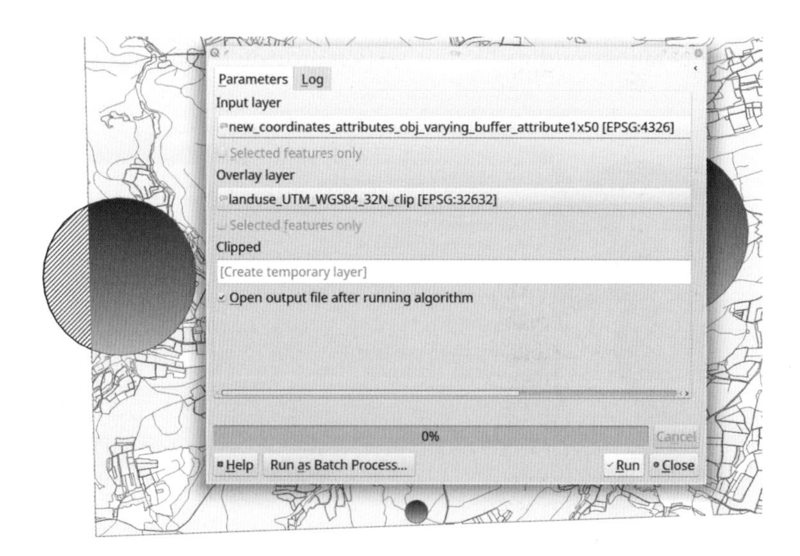

Figure 8.6 Clip function applied to location buffers of varying sizes and the land use vector object used as the overlay layer. The output and the original shape of the buffers are shown in the background. The buffer with the green gradient is the resulting object.

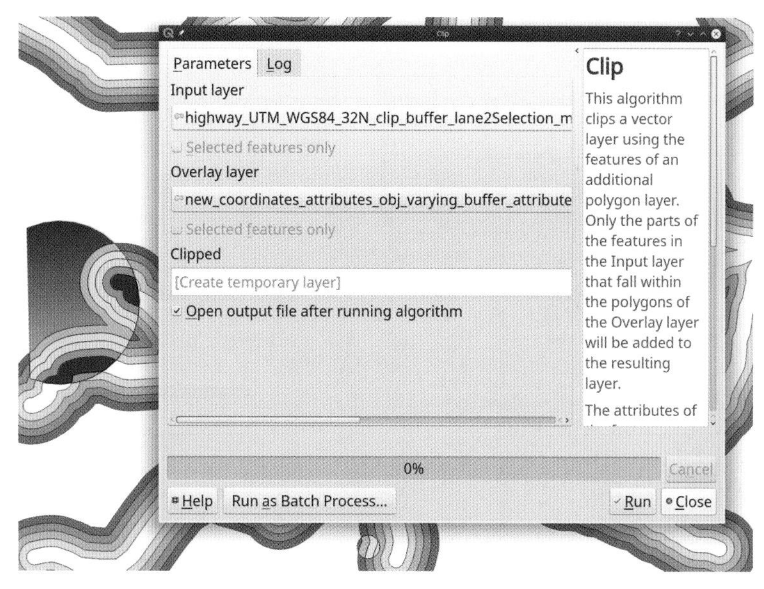

Figure 8.7 Road multibuffers clipped to the shape of the previously clipped point buffers. The original road buffer object is shown in the background in grey; the point buffer is shown in green. The resulting clipped road multibuffers are coloured from red to yellow and green.

```
Vector > Geoprocessing Tools > Clip ...
```

The same needs to be done for the multibuffer of the two-lane road object, but this time not to the extent of the study area, rather to the clipped buffer areas (Figure 8.7). In this case, the input layer is the multibuffer of the two-lane road layer, while the overlay layer is the output of our previous clipping operations. If you open the attribute table of the newly clipped object, you can see that one column remains, which contains the information of the different buffers. If we now combine (*union*) the clipped point buffer and the clipped road multibuffer, we get a much more extensive attribute table:

```
Vector > Geoprocessing Tools > Union ...
```

The spatial object is now a combination of the clipped road and point buffer object, including their attribute table entries (Figure 8.8). Single columns within the attribute table can be removed if we do not need them anymore by opening the attribute table, activating the editing and selecting the *delete field* function.

Now we have prepared our spatial object for the actual calculations. Having all spatial features within one object is, in our case, a prerequisite because we want to analyse the ratios between different polygons within each site. The required steps are:

- Add geometry attributes, namely *area*, to the attribute table for each individual polygon.
- Calculate the sum of all individual polygons grouped by site name.
- Calculate the ratio of total area to the area of single polygons for each site separately.

These operations can be performed within the attribute table by using the *Field Calculator* function (*slide rule* icon). Within the *Field Calculator*, a new or existing field can

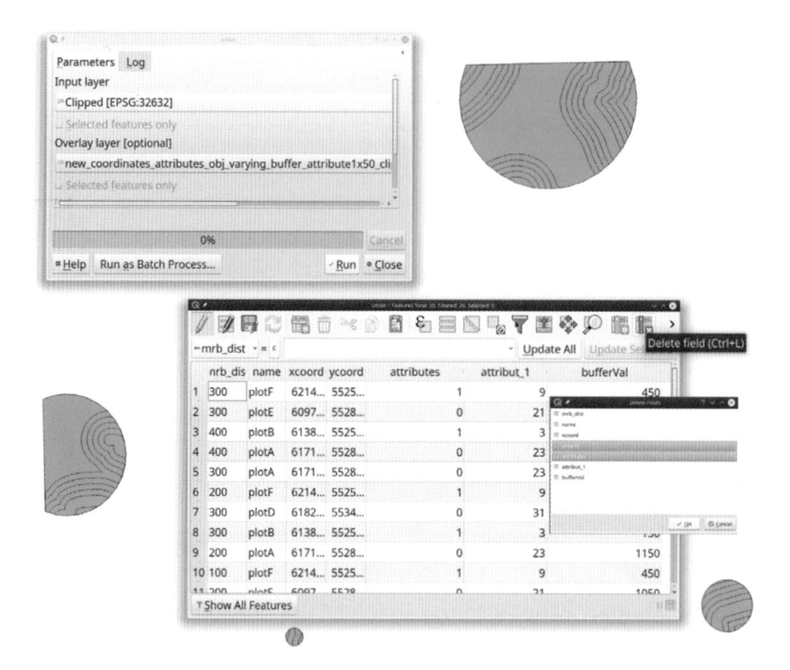

Figure 8.8 The clipped road and point buffer is combined (union), which results in a combined attribute table. Removing single columns is possible after activating the editing and selecting the delete field function.

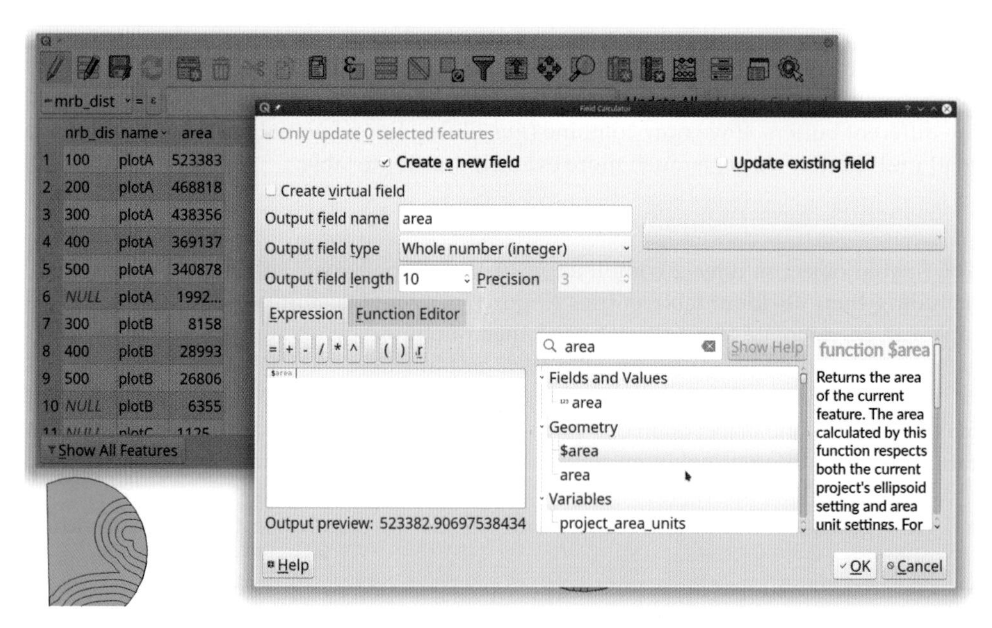

Figure 8.9 Within the Field Calculator, various calculations can be performed, such as common mathematical operations as well as geometrical analysis. In this example, we added a column named 'area' and inserted the area value for each polygon in our vector object.

be defined as well as the type of output and its length. Operations can be picked from a list and added to the actual calculator field (Figure 8.9). First, we create a new column in our attribute table that holds the area information for each single polygon. Next, we use the same interface but create another field, *sumArea*, and assign the sum of all area values grouped by the name column using the command:

```
sum(area, name)
```

This results in the same value for all polygons within the same site. Note that so far, all values are whole (integers) and not decimal numbers (floats); length is set to a maximum of ten entries. These settings must be customized based on your data and results. Calculating the percentage of the single polygon area and the total area of our point buffer requires decimal numbers. This needs to be changed before running the following calculation for the new field 'ratio':

```
"sumArea"/"area"
```

To display the percentage road area within each point buffer, as a final analysis, we calculate the sum of all ratio values, again grouped by site name, like the earlier calculation. Selecting one of the columns created within our attribute table results in either the total area of roads within each point buffer or the percentage of single multibuffer polygons (Figure 8.10). The last issue is that the lines of the multibuffer areas within the point buffers are still visible. These single polygons can be selected individually and queried, but we can also merge all polygons and get rid of separate polygons by:

```
Vector > Geoprocessing Tools > Dissolve ...
```

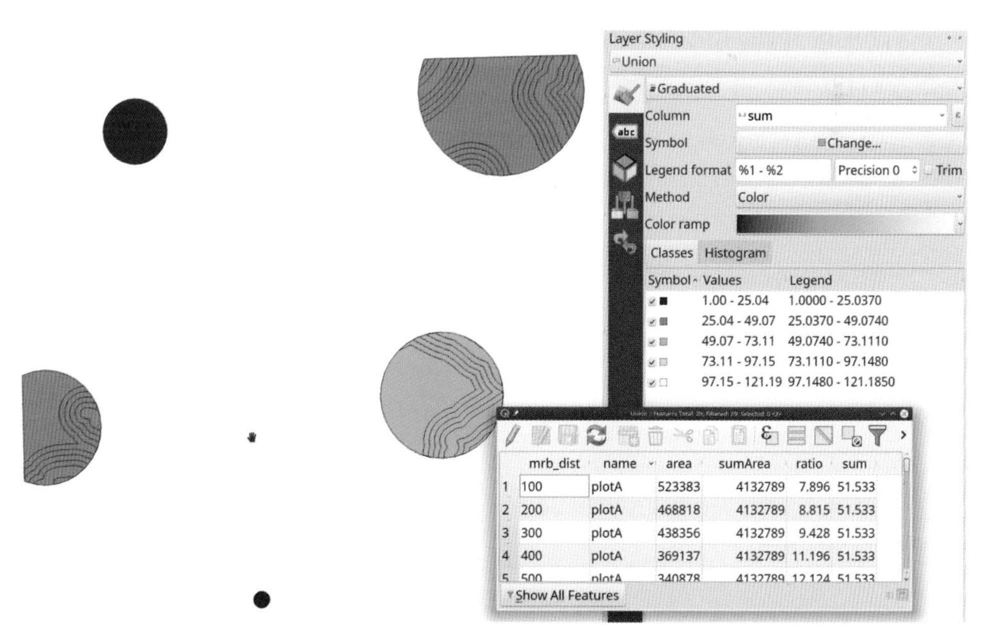

Figure 8.10 The resulting attribute table values can be opened and visualized. Note that values can be visualized for each buffer or segment depending on the columns selected; however, the segment borders will still be visible.

It is important to define the *dissolve fields* in this operation; otherwise, all polygons, regardless of which site they belong to, are dissolved. Therefore, we set the *name* column for this operation; the results are six individual polygons for site A–F, including their attributes.

This example made use of relatively simple field calculations. However, the Field Calculator is much more powerful. *If* statements to execute certain analyses can be added if predefined conditions are met. Moreover, conversion operations, fuzzy matching or coordinate reference system queries can also be chosen. To make it more powerful and complex, other spatial objects (*Map Layers*) or other geometry information (*Geometry*) can also be included in such calculations.

8.2 Spatial distances

A common task for a variety of analyses is the computation of distances between features. Distance information can be retrieved from a common distance matrix where single distances between individual features are provided:

```
Vector > Analysis Tools > Distance Matrix ...
```

Within this function, distances within a one-point layer between all individual points or distances between two different point layers can be computed, as shown in Figure 8.11. In both cases, a distance value for each pairwise connection is provided and stored in a new vector object, including an attribute column holding the distance information for each point combination. Such information can be used for further spatial analysis as well as exploring the visual display of distances between points (*Layer Styling* > *Graduated* > *Method: Size*). Within the same section, the *Nearest Neighbor Analysis* creates a non-spatial summary statistics file. However, these operations can only be applied to point layers and do not provide spatially explicit results. One approach to obtain distance information for

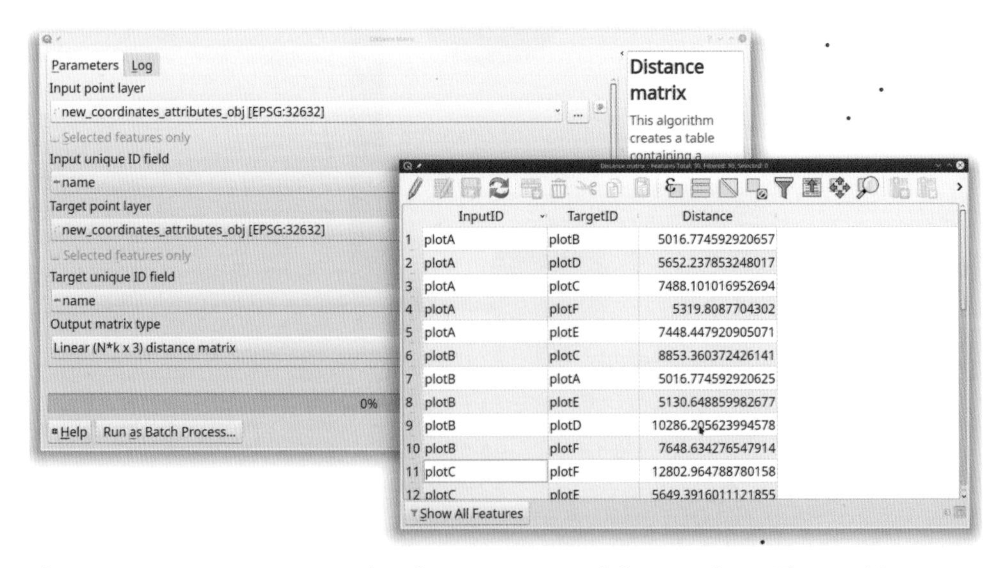

Figure 8.11 Distance matrix analysis between points of the same layer. The resulting vector layer provides the pairwise distance information within its attribute table.

points, lines or polygons is to use multibuffers, as shown earlier; an alternative operation would be a distance raster layer where each raster cell holds the distance value for the chosen feature. The latter approach bridges the vector-raster system.

Let us assume we want to calculate the distances of our field sites to all settlements or certain land use types in our study area. The approach would be the same for points, polygons or lines, such as places or roads. The *landuse_UTM_WGS84_32N_clip* polygon data set is used, and certain land use types are selected using the *Select By Expression* function within the attribute table (the icon showing an epsilon covering a yellow square). With this function, various attribute values can be selected using different operators. We use a simple *OR* operation and tell QGIS to select all polygons having the attribute 'farmland', 'residential' or 'industrial' in the 'landuse' column (Figure 8.12):

```
"landuse" = 'industrial' OR
"landuse" = 'farmland' OR
"landuse" = 'industrial'
```

Further conditions could be inserted, as well as other column attributes. For example, polygons that have the attribute 'farmland' in the column 'landuse' and (AND) any category in another column, for example, 'osm_type' should be selected. Also, larger than (>) or smaller than or equal to (≤) can be used, as well as further conditional operators. Regarding the current example, at the bottom right of the interface, the expression can be applied to the vector object and all matching features are selected. The selected polygons are highlighted within the attribute table and on the map view. If necessary, single polygons can be added or removed from the selection by using the *Select Features by area or single click* or by clicking on the *Attributes* toolbar and holding the Ctrl key while (un)selecting polygons. Otherwise, only a single polygon will be selected, and

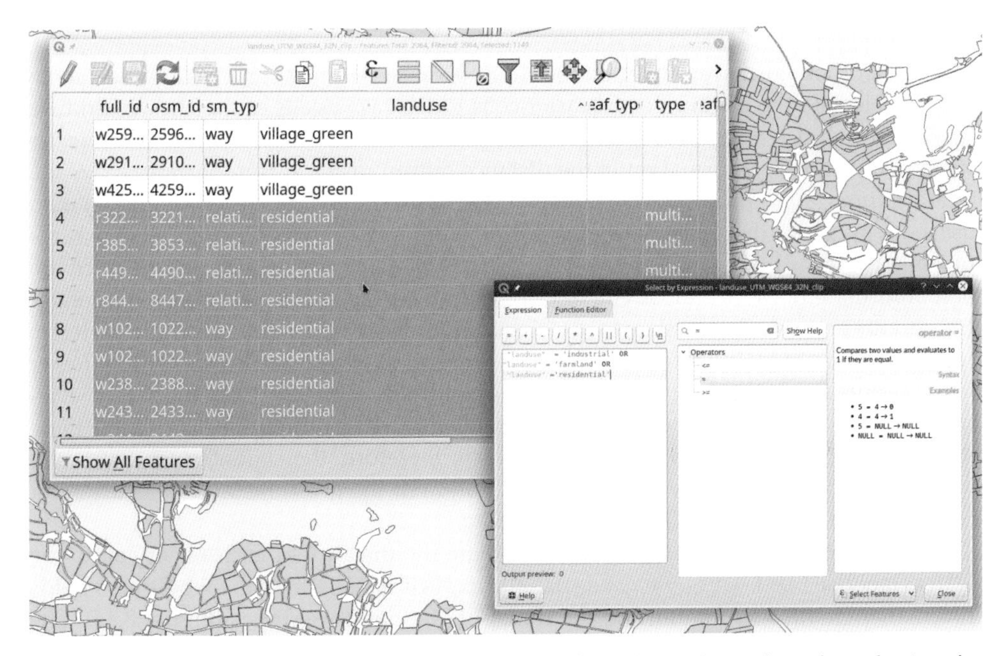

Figure 8.12 All land use that is either farmland, residential or industrial is selected using the Select By Expression function within the attribute table of the 'landuse' polygon vector.

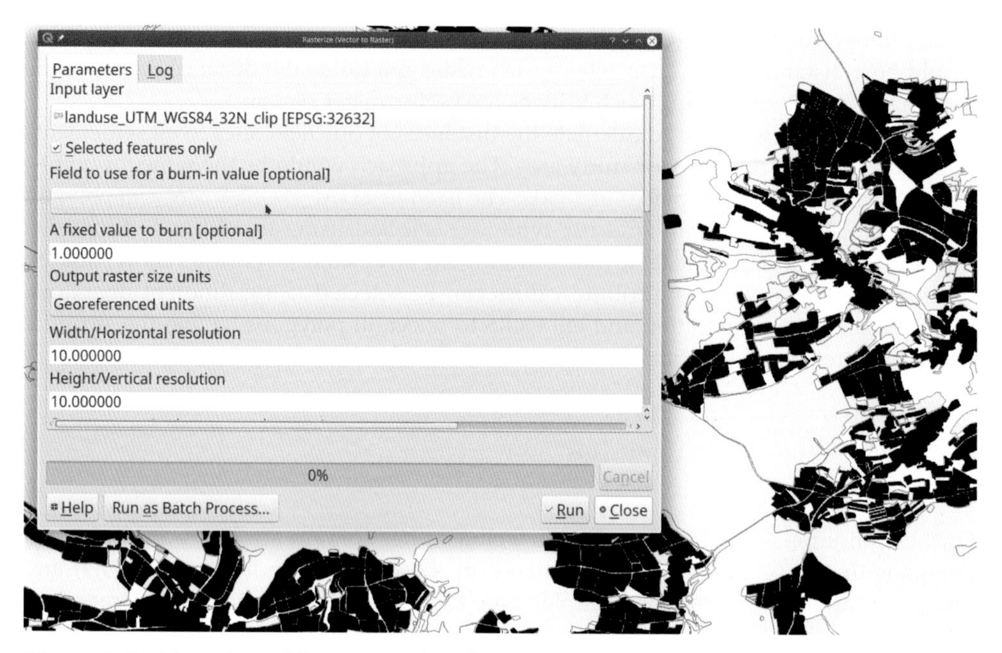

Figure 8.13 The selected features within the 'landuse' vector layer area are converted to a raster with a 10-m spatial resolution. The rasterized output and the original objects are shown in the background.

the previous selection will be lost. These selected features are then converted to a raster object through:

```
Raster > Conversions > Rasterize (Vector to Raster)...
```

The selected polygons are converted to a 10-m pixel size raster (Figure 8.13) and can be used to calculate the distance of all raster cells to the raster cells of the chosen landuse types. The distance units were changed from *pixel coordinates* to *georeferenced coordinates* to have actual metre information in the distance output raster. It is important to tick the *selected features only* option, otherwise all polygons will be converted to a raster rather than just those polygons that match our expression. The resulting raster layer holds the same information regarding spatial extent as the landuse vector object, but not the attribute table information. Moreover, if you zoom in you will see single raster cells, no longer the smooth vector polygon shapes. By changing the spatial resolution in the rasterize operation, you can make the pixel size finer or coarser. A higher spatial resolution (lower value) results in more accurate spatial matches with the original vector object. However, making it too fine increases computational time and the resulting file size. Choosing a very coarse spatial resolution impacts accuracy; therefore, the spatial resolution must be chosen with care. Usually, the spatial resolution of the given raster data is used, for example, 10 or 30 m for Sentinel or Landsat data.

The 'landuse' raster can be used to calculate the distance of each raster cell to the nearest landuse pixel. Thus, we get a spatially explicit result where the whole area holds information about its proximity to the three selected land use types (Figure 8.14). The distances are available as georeferenced coordinate units, but only in steps of 10 m because of the chosen spatial resolution during the rasterization process.

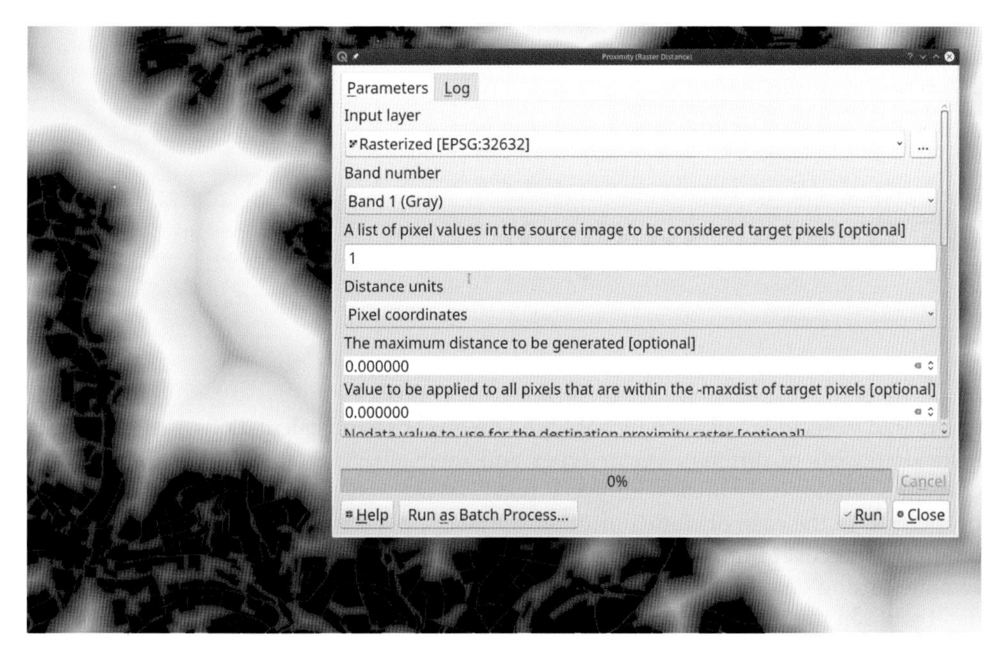

Figure 8.14 Distance values in raster format based on the 'landuse' raster. Distances are available in 10-m intervals because of the 10-m spatial resolution chosen during the rasterize operation.

TASK

Calculate a distance raster to other features, such as roads or settlements. Also calculate the distance to the forest edge but inside the forest.

8.3 Summary and further analyses

QGIS offers powerful tools to handle, transform and analyse vector data. Several vector conversion operations, such as polygon to line conversion, are valuable and can be accessed from the *Geometry Tools* menu. Sometimes, attributes must be linked to another object covering the same spatial extent. This *spatial join* can be achieved through joining attributes by location within the *Data Management Tools* section. Browsing or searching functions within the processing toolbox leads to many more potential vector modification options. Further analyses are feasible by using additional plug-ins such as *LecoS (Landscape Ecology Statistics)*, *HotSpotAnalysis* or *Visibility Analysis*. The *MMQGIS* plug-in also provides several more relevant functions, such as Voronoi diagram analysis or geocoding. The best option is to explore the different plug-ins and their functions to gain a good overview of their potential and limitations.

9. Raster analysis

Compared to vector analysis, which is based on pixel or band calculations, raster analysis instead consists of different vector types and their attribute table. Raster data always cover the whole study area, with every pixel having a value related to the land cover properties. In the case of clouds or other obstacles, pixel values might not provide the required land cover information and may have to be ignored in further analysis. Missing remote sensing data because of features such as clouds is a common issue and must be dealt with. In the following sections, we calculate spectral and topographic indices and create a categorical land cover map.

9.1 Spectral landscape indices

One of the most frequently used spectral indices is the *normalized difference vegetation index* (NDVI), which takes advantage of an increase in the near-infrared (NIR) if the vegetation is dense and photosynthetically active (Figure 9.1). Spectral bands of data provided by Sentinel or Landsat cover the red and NIR spectrum, which allows us to calculate the magnitude of increased reflectance. The NDVI, as well as many other spectral indices, can be used to calculate the ratio between red and NIR reflectance. The resulting NDVI

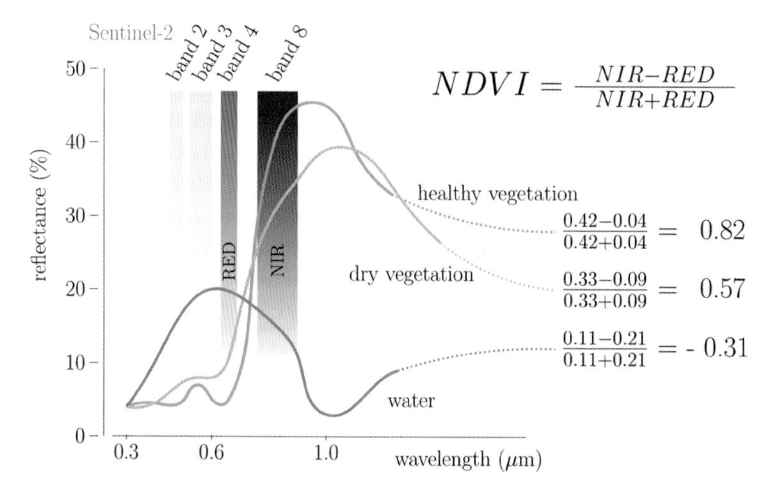

Figure 9.1 Spectral response of different landscape features and corresponding Sentinel band positions on the spectral wavelength. NDVI values are provided for each of the sample features. NDVI, normalized difference vegetation index; NIR, near-infrared.

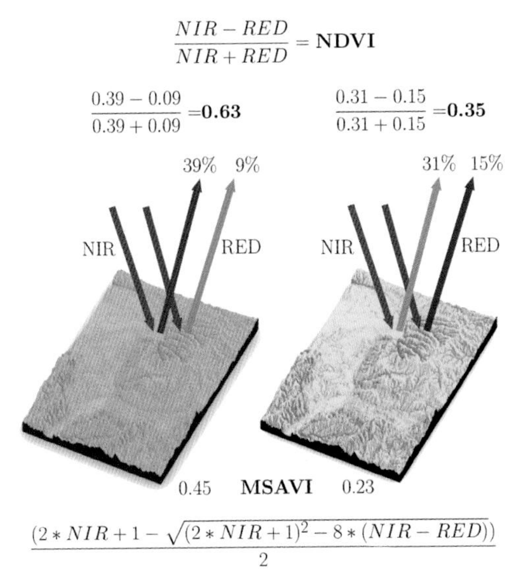

Figure 9.2 The NDVI and MSAVI formula and general values for a photosynthetic active and a less green landscape are shown. MSAVI, modified soil-adjusted vegetation index; NDVI, normalized difference vegetation index; NIR, near-infrared; SWIR, short-wavelength infrared.

values range between –1 and +1. Healthy vegetation has a higher value, approaching 1, whereas dry vegetation has a lower value, with bare soil close to 0 and water having a negative value. The NDVI, and mathematically more complex indices, such as the *modified soil-adjusted vegetation index*, generally aim to provide a proxy for vegetation quality or landscape condition (Figure 9.2). Other indices differ in their aim; they use different formulas and bands (Figure 9.3). The resulting raster maps provide different

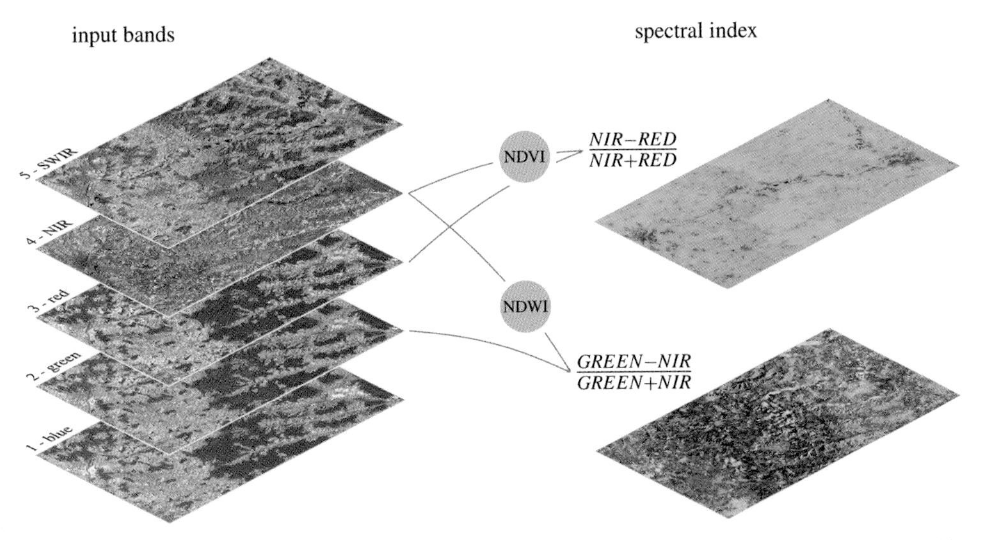

Figure 9.3 Indices and formulas used for different spectral bands to create valuable information for a specific purpose. NDVI, normalized difference vegetation index; NDWI, normalized difference water index; NIR, near-infrared.

Figure 9.4 Changing band combinations in the formula results in different landscape information and is addressed by different indices. NDBI, normal difference built-up index; NDVI, normalized difference vegetation index; NDWI, normalized difference water index; NIR, near-infrared; SWIR, short-wavelength infrared.

information and can be used for a variety of applications, from vegetation to water or snow analysis. Calculating a range of different indices for the same study area provides different insights into the properties of the landscape and can be visualized via the index values within each raster (Figure 9.4). Several indices exist, with the NDVI being the most prominent one. However, we highly recommended that you also evaluate other indices, especially for very dense or sparsely vegetated areas where the NDVI is thought to be less accurate. A list of indices is shown in Table 9.1. However, not all indices result in the same value range. Thus, every index and its formula(s) must be critically evaluated and the associated literature should be read.

The diversity of spectral indices is very high. Only a small subset is listed in Table 9.1. Historical indices tend to be simpler whereas newer indices have more complex formulas.

R COMMANDS

`calc()` or `overlay()` functions offer the same functions as the QGIS raster calculator using a similar syntax.

The `spectralIndices()` command within the RStoolbox package allows the calculation of various indices without entering any formula.

As a first step, calculating an index requires to load the corresponding spectral bands as raster layers. Thus, we load the Sentinel scene we have downloaded previously (T32UPA_A015792_20180701T102404_STACK_crop) into QGIS. We could also add the single-band raster layer, although the output would be the same. Applying a formula on the different bands is done with the Raster Calculator:

```
Raster > Raster Calculator ...
```

Within the *Raster Calculator*, bands and operators can be combined and the NDVI formula using the correct bands can be added. In Figure 9.5, the interface with the NDVI formula for our Sentinel stack is shown. The formula we enter is:

Table 9.1 List of known and commonly used indices

Acronym	Index name	Formula	Main properties	Comments	Reference(s)
RVI	ratio vegetation index	NIR/RED	One of the very first vegetation indices	Rarely applied nowadays	First described by Jordan (1969)
DVI	difference vegetation index	NIR–RED	One of the very first vegetation indices	Rarely applied nowadays	Lillesand and Kiefer (1987); Tucker (1979)
NDVI	normalized difference vegetation index	(NIR–RED)/(NIR+RED)	Responds to the amount of chlorophyll present	Values deviate when vegetation cover <30% and when vegetation is very dense	Rouse et al. (1974)
EVI	enhanced vegetation index	G×(NIR–RED)/(NIR+C1×RED–C2×BLUE+L)	More sensitive to differences in plant canopy; similar to the leaf area index; adopted as standard MODIS product; default values: L=1, C^1=6, C^2=7.5 and G=2.5	Mostly used to assess biomass and biophysical properties like the leaf area index, and to quantify evapotranspiration or water-use efficiency	Huete et al. (2002)
NDWI	normalized difference water index	(Green–NIR)/(Green+NIR)	Related to the status of vegetation liquid water	Applied to water, as well as in vegetation and public health studies	Estallo et al. (2012)
MNDWI	modified NDWI	(Green–MIR)/(Green+MIR)	Can enhance open-water features by removing noise from built-up areas	Used to extract water information for a water region against a background dominated by built-up land areas	Xu (2006)
SAVI	soil-adjusted vegetation index	(NIR–RED)/(NIR+RED+L)×(1+L)	Adjusts for soil effect on the vegetation index in instances of low vegetation cover; L depends on the soil background signal (default 0.5)	Semi-natural grassland mapping with adjustment for soil reflectance	Huete (1988)
MSAVI	modified SAVI	(2×NIR+1–SQRT($(2\times NIR+2)^2$–8×(NIR–RED))/2	Applies an additional soil brightness correction factor	Works better with lower vegetation cover or very dense vegetation cover	Qi et al. (1994)

MIR, mid-infrared; MODIS, Moderate Resolution Imaging Spectroradiometer; NIR, near-infrared.

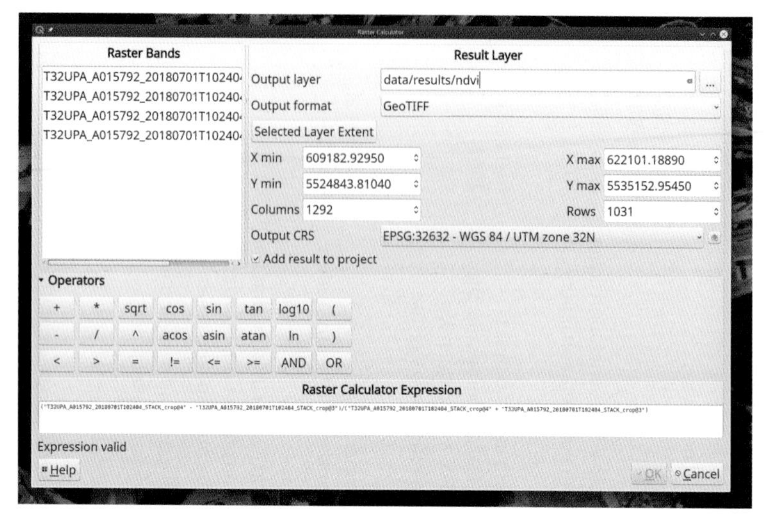

Figure 9.5 The Raster Calculator allows you to apply different operators on raster layers, such as the NDVI formula.

```
("T32UPA_A015792_20180701T102404_STACK_crop@4" - "T32UPA_
A015792_20180701T102404_STACK_crop@3") /

("T32UPA_A015792_20180701T102404_STACK_crop@4" +
"T32UPA_A015792_20180701T102404_STACK_crop@3")
```

In the formula, we replace the NIR and RED variables with the band names. It is always good to check the resulting image for the correct range of values, which should be between –1 and +1. Sometimes, artefacts might lead to higher or lower values for a few individual pixels, but these can be ignored. Displaying the NDVI raster with an appropriate colour scale and comparing it to the input data allows a first evaluation of the analysis (Figure 9.6). In Figure 9.6, the *Map Swipe Tool* has been used to swipe between two layers and compare the patterns. Dark green areas in the NDVI indicate high NDVI values, that is, dense photosynthetically active vegetation, while bare soil or less vegetated areas are shown in light brown. The legend in the *Layers Panel* and the raster histogram in the *Layer Properties* settings show an adequate range for our NDVI layer. The values range between –0.16 and 0.93, with a frequency above 0.8 (Figure 9.7). This frequency makes sense considering the high percentage of forest in the study area. Selecting individual pixels with the *Identify features* button (the blue circle with an 'i') also shows that pixels located in forest have higher NDVI values than those located in agricultural areas. However, some fields also have very high NDVI values; thus, we cannot deduce that high values are always indicative of forest cover. A high NDVI value simply indicates that this area has a dense photosynthetic vegetation cover. This issue must always be kept in mind. An index does not make a sound ecological knowledge of the study area redundant. Results must always be put into perspective and matched to known landscape history and cover. Using a series of NDVI values for different seasons leads to more in-depth information of landscape properties because forest types, as well as managed lands, change over time in a very specific manner. This is known as phenology and remote sensing time series analysis. A simple approach to incorporate the temporal dimension in such an analysis,

Figure 9.6 The resulting NDVI, shown as a single-band pseudocolour, and the RGB shown as a multiband real colour composite shown side by side using the Map Swipe Tool.

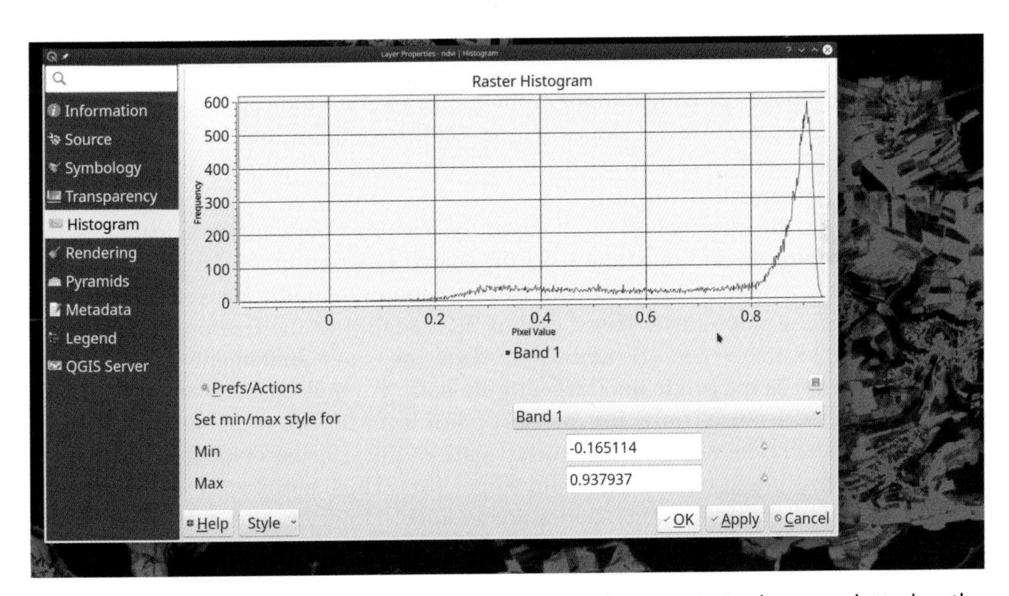

Figure 9.7 Histogram of the NDVI values within our study area. NDVI values are plotted on the x-axis, the number of pixels on the y-axis. This histogram shows that our study area has many high NDVI pixel values.

is to calculate the NDVI for several time steps per year and then calculate the mean and standard deviation across all raster layers. The resulting raster will hold information about the variability or average for each pixel in the landscape.

Calculate further vegetation indices, such as the MSAVI or global environmental monitoring index. Compare the resulting raster data sets visually with the *Map Swipe Tool* and calculate the differences using the *Raster Calculator*.

9.2 Topographic indices

Not only spectral bands can be used to calculate landscape information. The digital elevation model (DEM) provided by the Shuttle Radar Topography Mission (SRTM) data set can be used to derive relevant landscape information, such as slope or aspect. Additionally, terrain roughness or ruggedness can be calculated; this is the largest difference to the value of the centre pixel and the mean difference of the centre pixel and its surrounding, respectively. These functionalities and others can be accessed through:

```
Raster > Analysis > Slope ...
Raster > Analysis > Roughness ...
Raster > Analysis > Terrain Ruggedness Index (TRI)...
```

The resulting image provides topographic information, indicating areas of high elevational differences or steep slopes (Figure 9.8). Depending on the research goals, such topographic indices can be used for defining habitat variability, accessibility, or suitability for settlements or farmland.

For example, the information can be used for least cost path analysis. Installing the *Least Cost Path* plug-in and opening it via the search function in the *Processing Toolbox* allows us to calculate the 'cheapest' path from one point to another based on a friction raster. We take the slope raster based on our SRTM DEM as the friction or cost raster and the field locations (*new_coordinates_attributes_obj*) and settlements (*place_UTM_WGS84_32N_clip*) as the origin and target of the path. We must select one field location and tick 'selected features only' for the start-point layer but leave the end-point layer as it is (Figure 9.9). The result is the shortest path of the selected field site point to one of the settlements based on the minimum slope costs. Thus, paths with shallow slopes are preferred and areas with high slopes are avoided. However, this analysis does not consider the road network; thus, it might be useful for finding the best (low-cost) hiking routes or for checking the connectivity of two locations for animals.

9.3 Spectral landscape categories

A related approach, but one with quite a different output, is the method used to derive discrete classes within a landscape. This method is called *classification* and is commonly applied to general land cover, settlement or agriculture mapping. When no previous knowledge about land cover types (classes) is available, it is called *unsupervised classification*. Spectral band values are assigned based on statistical methods, for example,

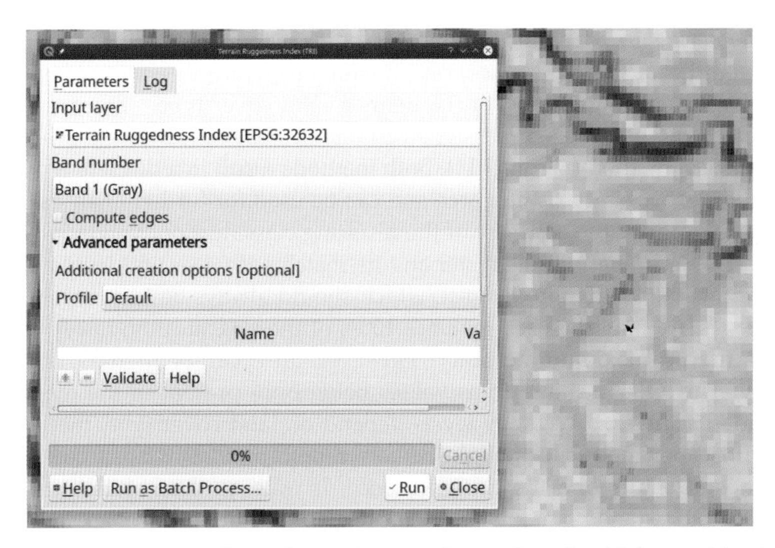

Figure 9.8 Various topographic information can be retrieved, which provides landscape information related to, for example, elevational differences. One interface is displayed with its corresponding output raster shown in single-band pseudocolour.

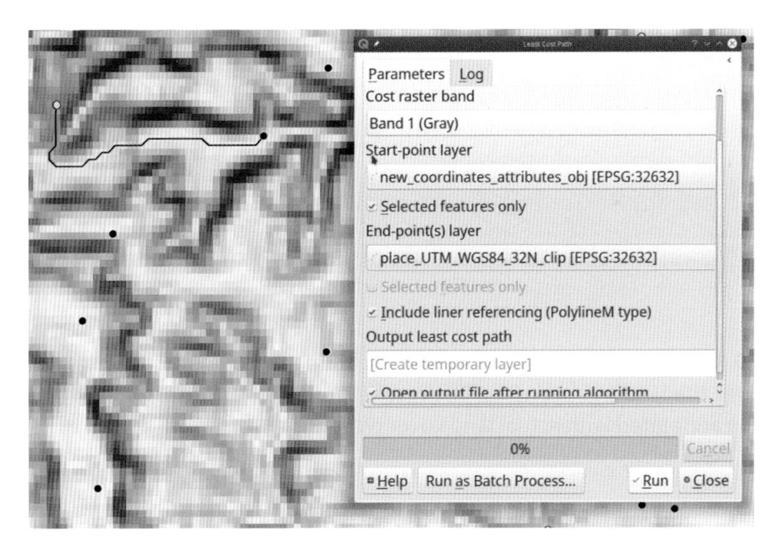

Figure 9.9 Least cost path analysis between a selected field site location and all settlements based on a cost raster. The cost raster in this example is the slope, assuming steep slopes are not preferred for hiking, for example. The least cost path is shown in the background together with the slope raster values (the areas shown in dark red hold a high slope value).

k-means clustering to a defined number of homogeneous groups (Figure 9.10). The opposite would be a *supervised classification*, which provides the spectral bands, as well as a vector with training classes, of the defined land cover types. These two data sets are combined, and all pixels are assigned to one of the predefined classes based on a chosen statistical method (Figure 9.11). Both methods result in a discrete representation of the land cover within our study area and can be used for further analysis. Validation is required to evaluate the accuracy of the resulting classification. Validation checks the resulting class assignment in the raster and its match with a second validation data set (Figure 9.12). If the validation data set, which should always be kept separate and not mixed with the training data set, matches 100% of the classification raster values, then we have achieved maximum accuracy. However, an accuracy of 100% is rather rare or unlikely. This would show up as values being only present in the diagonal fields of the validation matrix. Accuracy levels depend on land cover type, the legend used and the landscape; good values might be between 0.7 and 0.9. Also, accuracy depends on landscape structure and the classes chosen. If water and no water are used for a classification and water bodies are quite clearly distinguishable, then very high accuracy can be expected (>0.9). However, a diverse landscape with gradual transitions and many land cover classes will naturally result in lower classification accuracy. Such a relationship must be considered when accuracies are assessed. Training and validation data sets can be kept in the same vector object and split internally (Figure 9.13) or two separate training and validation vector objects can be created. In the latter, validation polygons should be created before the actual classification, and should be distributed evenly and

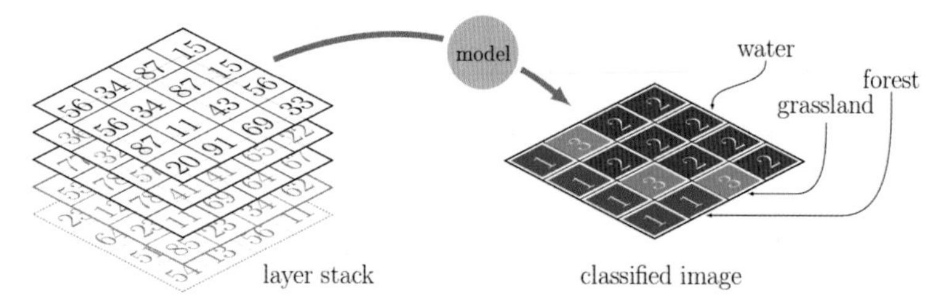

Figure 9.10 Classification approach using an unsupervised method where the statistical model defines the land cover classes. The resulting land cover classes are given a value and must then be annotated with actual land cover types.

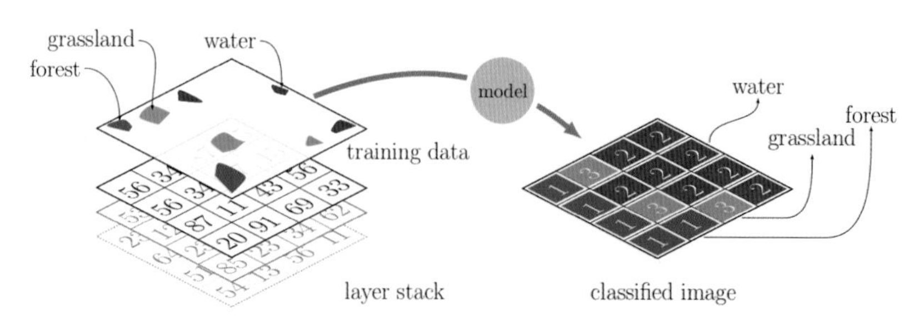

Figure 9.11 Supervised classification approach using existing class information provided by a vector object. The resulting classification is related to the input training data; thus, land cover types are already known in the output raster.

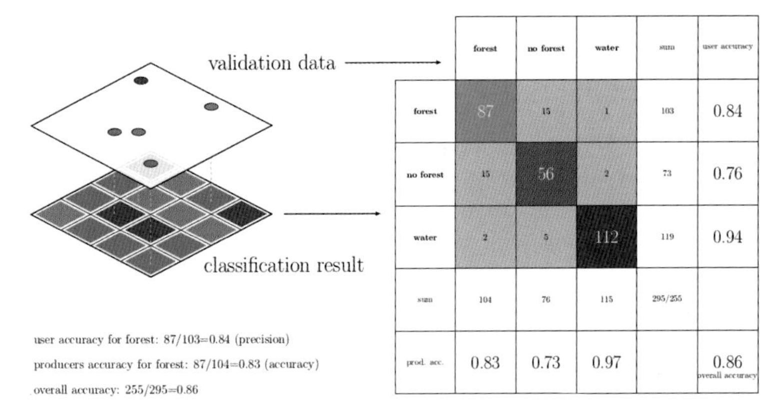

	forest	no forest	water	sum	user accuracy
forest	87	15	1	103	0.84
no forest	15	56	2	73	0.76
water	2	5	112	119	0.94
sum	104	76	115	295/255	
prod. acc.	0.83	0.73	0.97		0.86 overall accuracy

Figure 9.12 Graphical outline showing the validation of the results from a supervised classification using an independent validation data set, which are usually polygons but can also be points.

chosen objectively to avoid a bias towards certain landscape structures. This would lead to an artificially higher accuracy due to the researcher unintentionally selecting more suitable areas for validation. The optimal approach would be to place random or regular validation areas in the study area or use a second person for the validation. Validation polygons should cover homogeneous areas and should not cover different landscape elements, for example, forest and water bodies within one polygon.

Overall, validation outcomes should result in producer and user accuracy, among other measures. Overall accuracy is the average across all land cover classes. If the accuracy or classification error is unequally distributed, we would not be able to visualize it within the overall accuracy. However, looking at the validation matrix (Figure 9.12), not all values within the diagonal fields indicate reduced accuracy for that specific class.

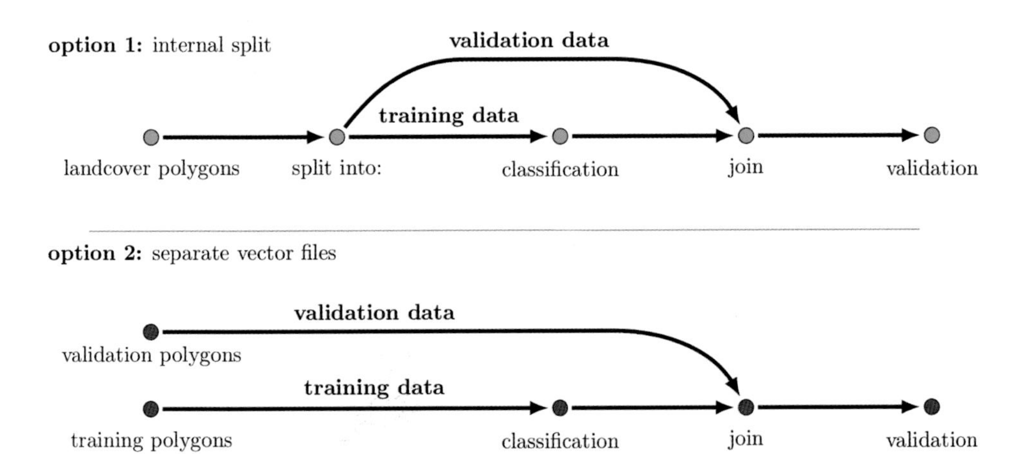

Figure 9.13 Validation approach using a separate validation vector object, which evaluates the matching classes with the result of a classification. The validation object can be a sample of the training data set or a fully independent data set.

User and producer accuracy should be used to overcome this shortcoming. User accuracy defines how many pixels are erroneously classified as a wrong class (commission error). Producer accuracy tells us the opposite, that is, the omission. For you, the user, user accuracy tells you how many of the classified pixels are in the correct class (number of correctly identified pixel/total number in that class). Producer accuracy tells us how many of the pixels in our classified raster are assigned correctly. Further validation measures, such as Cohen's *kappa*, can tell us about the actual agreement and the agreement expected by chance. In all cases, accuracy must be reported so that your audience can assess the quality of your classification approach.

In our example, overall accuracy is 0.86, which depends on the landscape structure being acceptable. The best result (1.0) could only be achieved if only the diagonal fields (green, brown and blue) included values; thus, classification and validation match by 100%, which is rarely the case. In our example, as expected, water has quite high accuracies, as seen in Figure 9.12, while the other two landscape classes typically have lower values. Increasing the number of classes or omitting the water class would naturally result in lower overall accuracy. Therefore, the quality or accuracy of your classification must always be interpreted regarding landscape structure and the number and type of classes included.

An unsupervised or supervised classification can be conducted in QGIS using the *Classification Tool* plug-in. Using this tool, the input raster data, training data and type of algorithm to be used for the classification can be selected (Figure 9.14). The Sentinel scene and vector object holding the training data samples, which needs to be in the same projection as the raster data set, are selected. The accuracy values are saved as two comma-separated value (CSV) files in the same directory as the output classification raster. Further details about classifications using R are provided in the Chapter 13.

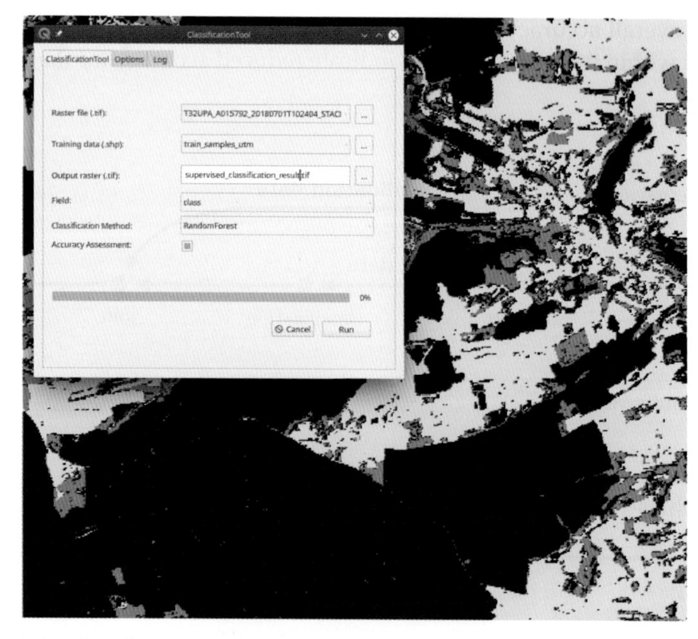

Figure 9.14 QGIS classification tool with the selected data sets and classification method. The resulting classification can be seen in the background. The accuracy values are saved in CSV format in the same folder as the classification output.

TASK

Create a new raster stack with the spectral bands of Sentinel and the SRTM DEM (see Chapter 3) and execute the classification again using this stack. Evaluate the differences between outputs.

9.4 Summary and further analysis

Raster data provide spatially explicit information about land surface properties. By downloading various scenes of the same area, a temporal analysis can be conducted and any changes in land cover can be detected. Acquiring two images from the same season but in different years allows us to calculate any differences. The simplest approach is subtracting one NDVI from another. More complex approaches are also feasible, such as *change vector analysis*. For changes in classification, a variety of methods can be applied. All these approaches and methods are addressed in detail in *Remote Sensing and GIS for Ecologists*. Additionally, methods such as land cover classifications might be feasible in QGIS using the *Semi-Automatic Classification Plugin* as well as the embedded functions of GRASS, Orfeo Toolbox or SAGA. We recommend searching for adequate functions related to specific research questions and then picking the right tool for the desired approach.

10. Raster-vector intersection

Combining raster and vector data is usually required for any analysis. Hence, intersecting two data sets to get the summary statistics of one layer within parts of another, such as polygons, is a core functionality. For example, knowing the normalized difference vegetation index (NDVI) or distance values for the locations of the in situ measurements or for certain areas is useful if we wish to analyse whether all land use polygons have similar NDVI values or whether our field locations have similar distances to the settlements. Doing this within QGIS is feasible through the *Point Sampling* and *Zonal Statistics* tools for points and polygons, respectively.

The approach used for different vector objects is partly different. Raster values behind points do not require any statistical aggregation since it is needed for vector polygons (Figure 10.1). For vector objects covering more than one pixel of a raster, not just one but many values represent this area. Therefore, summary statistics, such as the mean, maximum, minimum, range or standard deviation, must be applied. Each of the chosen statistical aggregation values are added to the vector attribute table as an individual column. Each polygon is assigned statistical values for the raster values that spatially correspond.

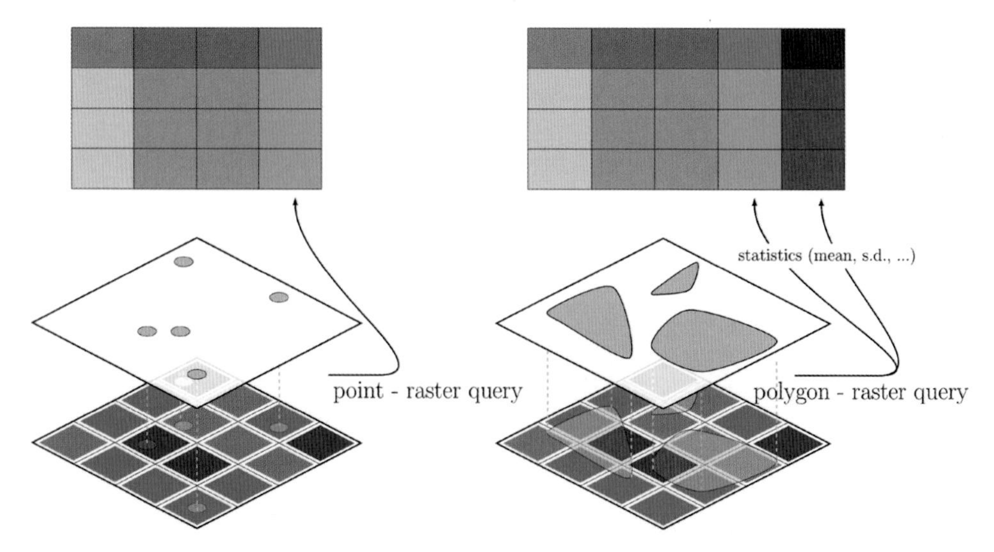

Figure 10.1 Raster values behind vector objects can be added to the vector attribute table. The statistics that reflect the raster values behind the vector objects must be defined for polygons.

10.1 Point statistics

For the point statistics of our field sample locations, we take the distances to settlements and annotate our points with the distance to the nearest village(s) at that location. The *Point Sampling Tool* plug-in allows us to define a point vector and layer to get the values from (Figure 10.2). It can be accessed through:

```
Plugins > Analyses > Point Sampling Tool ...
```

This function creates a new vector object, which holds the raster values; however, all other attributes are lost in the process. Therefore, we need to run the follow-on operation *Join Attributes by Location* tool to combine the attribute columns and assign them to the original point vector:

```
Vector > Data Management Tools > Join Attributes by
Location ...
```

Within this function, we can define when attributes should be joined. For our two-point vector object, it does not matter what we select; however, when using two different vector objects, using the *Geometric predicate* options can be very powerful. The same is true for the *Fields to add* and *Join type* options. We assign the distances to the original *new_coordinates_attributes_obj* point vector; then, several analyses or explorative graphical tasks can be carried out on that vector object using *DataPlotly*.

This analysis now allows us to combine field measurements and additional environmental variables, such as distance to settlements and NDVI values, and analyse them for our point locations.

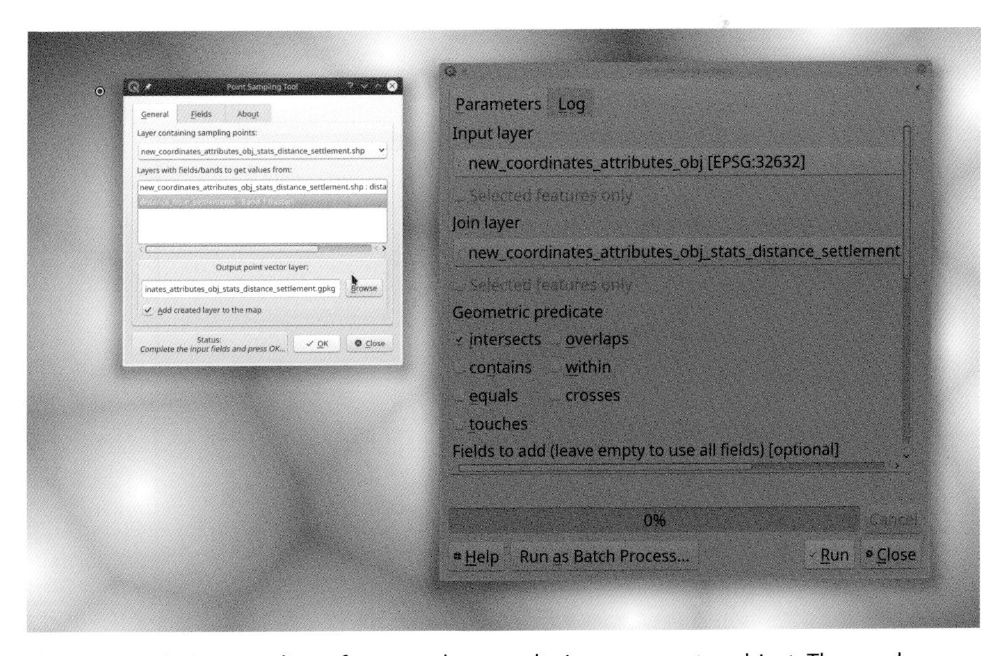

Figure 10.2 Point sampling of raster values results in a new vector object. These values can be reassigned to the original vector object through joining attributes by location. In this screenshot, both interfaces and distance raster are shown.

> **TASK**
>
> Add further environmental information to the point locations, for example, the NDVI values.

> **R COMMANDS**
>
> Using different R commands allows you to extract vector or raster data behind a spatially matching object:
>
> ```
> extract()
> over()
> ```

10.2 Zonal statistics

Searching in the *Processing Toolbox* (Ctrl + Alt + T) for 'zonal' retrieves two relevant functions, namely *Raster layer zonal statistics* for raster-raster statistics and *Zonal statistics* for raster-vector polygon statistics. We focus on the latter and aim to derive all NDVI values within the vector land use type 'residential' to check which settlements are the greenest.

We use the *landuse_UTM_WGS84_32N_clip* vector polygon data set and select all polygons with the land use attribute 'residential' via the *Select Features By Value* function (shortcut: F3). The selected polygons need to be saved as a new vector polygon through the normal export option (Right-click on *Layer* > *Export* > …), but this time we select the *Save only selected features* box, otherwise all polygons are saved (Figure 10.3). This newly

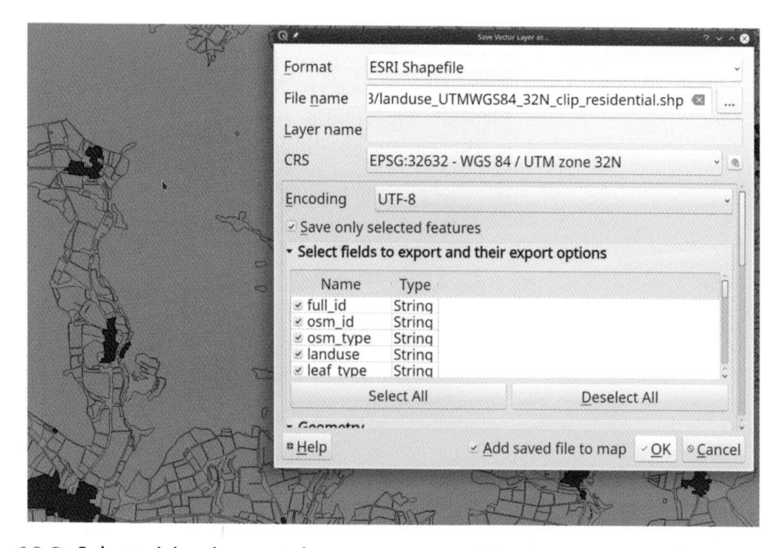

Figure 10.3 Selected land use polygons are saved in a separate polygon vector object. Selecting the Save only selected features box is important, otherwise all features would be saved.

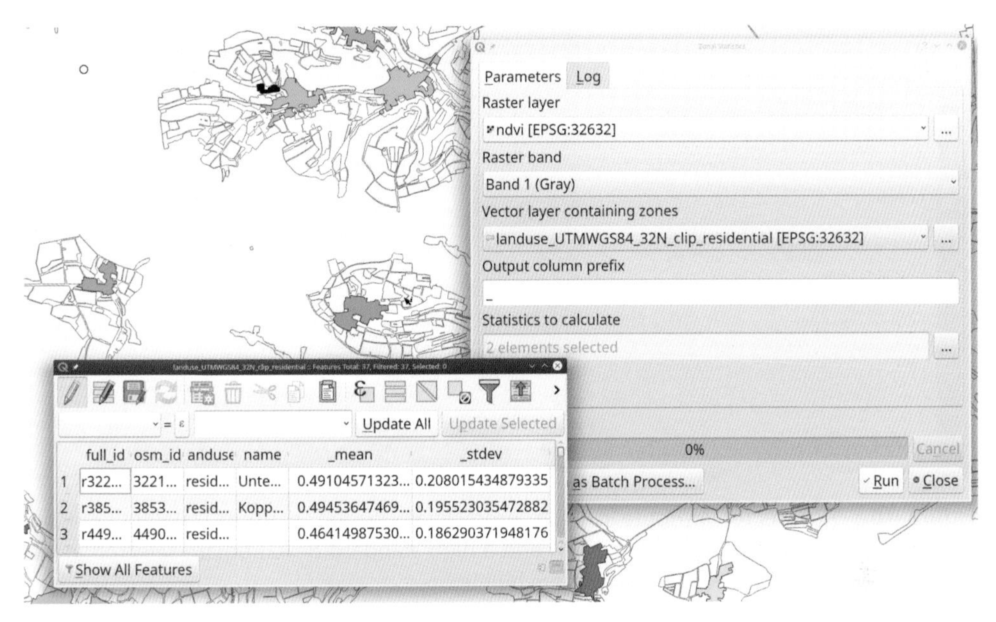

Figure 10.4 The Zonal Statistics function can be used to calculate defined statistical values for the polygons, based on an input raster. The resulting table and coloured vector file with the new statistical values are shown.

created vector object is then used within the *Zonal Statistics* algorithm as the input for the *Vector layer containing zones*. The NDVI raster layer is chosen as the input object to extract the values behind the vector polygons. Within the NDVI layer, we have a single band; if we selected a raster stack, then a single raster could be picked. The prefix for the name of the output column in the attribute table of our vector can be changed to something more meaningful, for example, 'NDVI_'.

To select the correct statistics for each polygon based on the NDVI, values must also be selected. In our example, we select the *mean* and *standard deviation* as the statistics (Figure 10.4). The resulting statistics for the selected residential polygons can be explored with the *DataPlotly* plug-in (Figure 10.5). Some residential areas have high NDVI values and low standard deviations within their extent, whereas others have a high variation of NDVI values within their area. This allows us to select residential areas where the NDVI is low and the standard deviation is high or vice versa for further analysis or for explorative purposes.

The resulting information, a new column in the vector attribute table, is like the point sampling performed previously; the only difference is that certain statistics are applied for the aggregation. Such information can also be used to analyse the attributes of an environmental variable within fields or study areas in subsequent statistical analyses.

TASK

Analyse and display distance versus NDVI values for land use or field sampling points. Explore different *DataPlotly* graphs, for example, contour and violin plots.

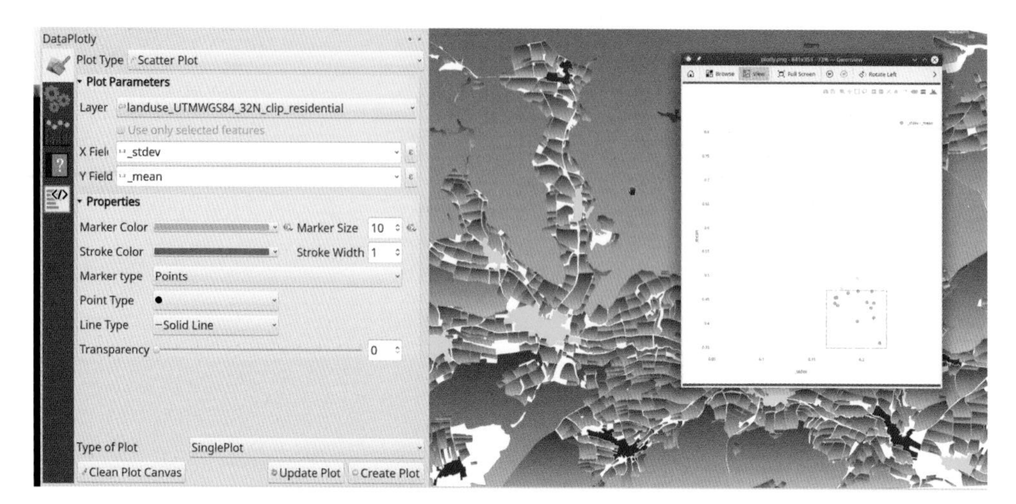

Figure 10.5 Any vector attribute can be interactively displayed using DataPlotly. In the screenshot, all polygons with a low NDVI and a high standard deviation are selected in the scatterplot; the corresponding polygons are highlighted.

The same procedure could be carried out for our field sample locations, the variable buffers of these locations or the road buffers. Such intersecting of different data sets and types is a common task and can be achieved using different methods and operations in any spatial software.

10.3 Summary

Intersecting raster and vector data, usually remote sensing and field data, is a common task but comes with various challenges, such as matching spatial or temporal data sets. This chapter has provided you with the basic knowledge of how to combine and analyse vector and raster data. Other, more sophisticated options exist and can be explored if your research aims require it.

Part IV

Spatial coding

In Part IV of the book, we introduce the R command line and perform some of the analyses we discussed in previous chapters again, but this time we use R commands (functions).

We introduce you to the world of coding in the spatial realm and highlight its advantages and potential benefits.

11. Introduction to coding

The previous chapters introduced you to spatial data handling using QGIS. Info boxes in those chapters provided you with some ideas of the corresponding R commands (functions), which will be introduced in detail in this part of the book. We assume you have not been exposed to programming in general, R or especially spatial coding. Covering the vast functionality of coding in general or within R is outside of the scope of this book. However, there are many books and online resources that provide further details on a variety of topics. We recommend two resources that introduce R:

- R for beginners (https://cran.r-project.org/doc/contrib/Paradis-rdebuts_en.pdf) by Emmanuel Paradis, the introductory text being a good starting point for everyone being new to R.
- Spatial data science in R (https://rspatial.org), which provides a helpful introduction to R for spatial analysis.

Since R and its community packages are developing quite rapidly, we highly recommend that you take some time each week to check for new developments, for example, by reading dedicated R blogs or following R developers on social networks such as Twitter, to be always informed about new features, book releases or even packages.

11.1 Why use the command line and what is 'R'?

In general, using a command line interface (CLI) is just a different way of interrogating a computer than using a graphical user interface (GUI). While a GUI uses windows, buttons and menus to get inputs from or return outputs to the user, a CLI expects the user to type a command to a console, which follows a certain syntax and returns outputs as text on the console or as files written to a directory. Thus, working with a CLI requires the user to know which commands are available and how to use them, whereas working with a GUI is usually self-explanatory. After getting used to it, a CLI usually allows you to communicate your computational tasks more easily and quickly than using a GUI. Working with a CLI has many advantages over doing the same tasks using a GUI:

- Your analysis is saved as a script and can be checked, modified and run any time.
- Informing others how a certain analysis has been performed can easily be achieved by sharing the script or parts of it (in contrast to explaining every single step in a GUI).

- Full scientific reproducibility and transparency of an analysis can be achieved by providing the script and the data sets or sources used by it.
- Designing and conducting an analysis using the command line is usually faster than using a GUI, since you get used to those commands you need quite often and there is no need to do every step manually by clicking.
- Analytical steps conducted in previous analysis that you have already written as code and that you want to apply in a new analysis can easily be transferred.
- Running repetitive tasks (e.g. doing the same operation repeatedly on different data sets) can be automatized, saving you a lot of time and effort.

R is a programming language and environment specifically designed for statistical computing and graphics. It is available from https://www.r-project.org for various operating systems. Over the last decade, R has become one of the most frequently used tools for both classic statistical and cutting-edge data analysis across many research disciplines. Although R started off as a statistical language, it has evolved into a multipurpose programming environment with comprehensive spatial data handling capabilities. R offers an incredible number of functionalities through so-called *packages*. These are developed and contributed by a large development community and cover all aspects of data manipulation, statistics, data mining and plotting across many scientific disciplines. Packages are a collection of functions that serve a certain purpose. A small number of packages are already installed when you use R for the first time. These so-called *core packages* provide all basic functionalities that other packages, developed by the community, build on. R does not come with a GUI like QGIS. Since R is operated using a command line, all you need is a text editor to write down and save your commands in script files. However, several programs, so-called *integrated development environments*

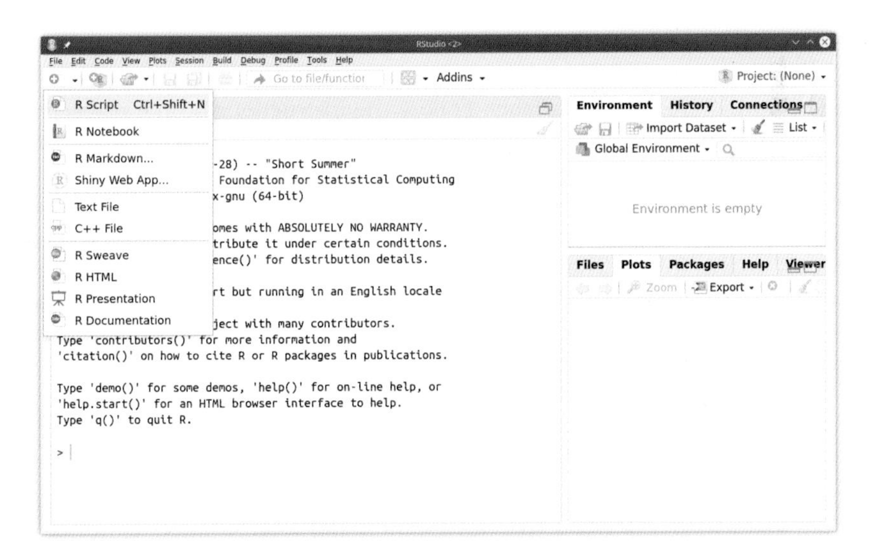

Figure 11.1 After launching RStudio, you will see three windows. The actual R session runs in the bottom left window; the top right window shows the environment; and the empty plot window is to the bottom right. To start a session, we need a fourth window where the actual script is written and where we can save our code in. Selecting the blank page with a *plus* sign or using the shortcut Ctrl + Shift + N opens a new R script.

(IDEs), provide convenient editors specialized in writing and executing R code; we highly recommend that you make use of one of these. Currently, the most popular and beginner-friendly IDE for R is RStudio (Figure 11.1) which you can download for free from https://rstudio.com.

In general, R does not assume to be more intelligent than you, the user; thus, it will not provide any suggestions or corrections. Rather, it provides warning and error messages instead. Entering a command that has a typo will result in an error message and make you highly accurate in how you write. Usually, an error or warning message will point you to the problem and solution; thus, you should make use of them and read them carefully. If you do not understand the message, just copy and paste it into any search engine. You are very likely to find a solution because you are probably not the first person to experience this problem.

11.2 Getting started

All primary data sets used throughout this book and the data sets produced by the analysis described in the book (called results or products) are provided. However, we intentionally decided not to provide the code of the analysis in this chapter because we believe that writing code (and experiencing error messages) is a key aspect of learning how to code. The code snippets shown in the following sections can be combined into a working script by you. If required, more advanced or lengthy scripts or updates on the code are provided on our Web page (http://book.ecosens.org), where all raster and vector files used throughout this book can also be found.

To start, you need to install:

- R from https://www.r-project.org.
- An IDE that provides a better interface for working with R. We use RStudio (https://rstudio.com) in the following sections.

R and RStudio are freely available. Follow the instructions on the respective Web pages on how to install them on your operating system. Please make sure to install R before RStudio.

The next step is to create a folder where you will store all scripts (normal text files with code in them). Within your file browser, you could create a text file called 'my_first_script.R' in this folder. The *R* suffix is telling your operating system that it is an R script and which software should be used to open it. For any later work, we advise using separate directories for different studies. For now, it might be best to have a series of scripts in your working directory that relate to the single chapters of this book.

11.3 Your very first command

Using R within an IDE like RStudio provides additional windows, such as a text editor for writing and saving the script. We use RStudio and type commands in the top left window. If there is none, please select the icon resembling a blank page with a *plus* sign and a new script will be opened (or type Ctrl + Shift + N). Several script writing windows can be opened and used simultaneously. If we want to execute commands from this window at the top left or rather send them to R, we can do so via Ctrl + Enter (or using the button 'run' located on the top right of the window; keyboard shortcuts are faster in the long run). We start with simple mathematical operation:

```
4 + 5
```

TASK

Try not to use the mouse or your track pad but just the keyboard. An overview of keyboard shortcuts in RStudio can be accessed via alt + Shift + K (or Help > Keyboard Shortcuts Help)

The result is shown on the screen but not saved for later use. This can be achieved by assigning the result to a variable:

```
x <- 4 + 5
```

You will notice that the result is not shown on the screen any longer; instead, it is saved to the *global environment* as 'x'. In RStudio, the *global environment* is displayed in the upper right corner by default, where you can check the content of a variable by clicking on it. We can also print the saved result by calling 'x' in the R terminal (Figure 11.2) or plot our result by typing:

```
plot(x)
```

Presently, the plot does not make sense, but you get the general idea and understand how simple the command line is. R offers a very large and much more complex command ecosystem that targets a variety of applications; of course, it takes time to master this ecosystem. If you do not know a command or want to read the help page of a specific command, you can search for it in the *Help* window on the right-hand side or hit F1 while your cursor is located within your command name. In general, make sure that you read the associated documentation when you use a new function (type either `?newfunction`

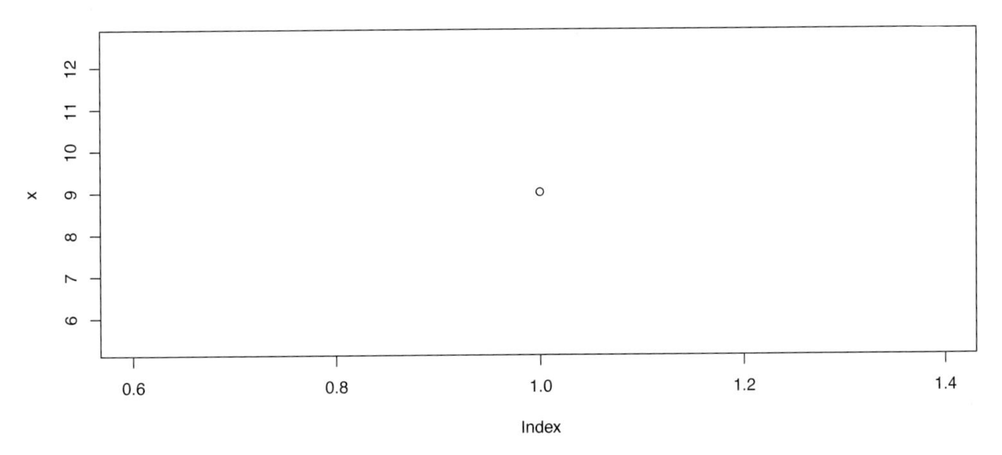

Figure 11.2 Simple plot of the values within the object 'x'.

or `help("newfunction")`). Most functions come with a set of default settings you should be aware of.

All help pages are structured in the same way, starting with a description, the command itself, its arguments and further details. The example section at the bottom of each help page is probably the most helpful to get started with a command. Just try the examples to understand the command and start from there: check the format of the input and how your data can fit into that structure, replace the example data with yours and run the analysis.

In the following sections, several key functions are introduced to prepare you for working with spatial data in R.

11.4 Classes of data

The most basic class of data is a *vector*. This vector should *not* be mistaken with the spatial vector data type! In R, we talk about a list of values called *vector*, as in the following example:

```
v <- c(5, 6, 2, 7, 9)
```

We can create a longer vector that contains random normally distributed numbers with a mean of 0 and a standard deviation of 1 using the `rnorm()` function and plot it again (Figure 11.3):

```
v <- rnorm(50)
plot(v)
```

Modifying the plot layout can be achieved through various settings within the `plot()` function; please check the help pages for all the details. In the following chapters, we discuss non-spatial graphs and related packages and commands (functions) in more detail.

Next, we want to create a matrix, which is a more complex data class and can be considered a vector in rows and columns:

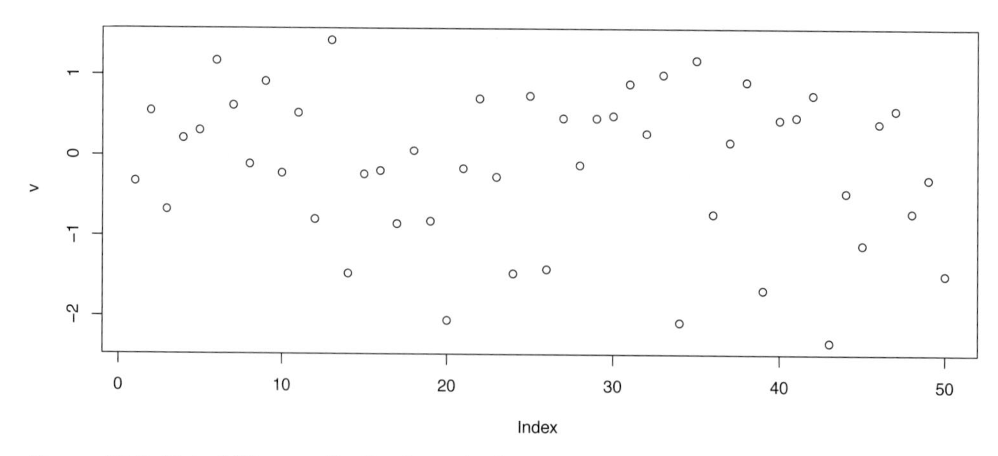

Figure 11.3 Plot of 50 normally distributed values.

```
m <- matrix(rnorm(15), ncol = 5, nrow = 3)
m

#> [,1] [,2] [,3] [,4] [,5]
#> [1,] -1.1502553-1.3969026 0.1653809-0.4278632-0.6413576
#> [2,] -0.2744712 0.7490577 1.1298091-0.2598021 0.1124575
#> [3,] 0.5779010-1.0511867 1.1737225-1.4111730 0.4226043
```

Looking at this data class, you might notice similarities to the raster data we have used before, that is, values in rows and columns. The only thing that is missing is the geographical information. We will cover this in the next chapter.

A similar data class is called *data.frame*. This is comparable to a spreadsheet table. It can hold not only numerical values, but also a mixture of different data types, such as *character* (text) and *numerical* (numbers) data. We can create a data frame by converting a matrix to a data frame or from scratch:

```
df <- data.frame(m)

df <- data.frame(colA = c(4, 6, 7, 3, 9), colB = c("a", "b",
"c", "d", "e"))
```

Now, we have a table consisting of a row with values (colA) and a row with characters (colB). This looks like the table associated with our spatial vector files, where we stored measurements and field site names in the same table.

The last class we want to address is the *list*. Lists can hold objects of different classes (e.g. the ones described earlier). For example, you can put a *vector, matrix* and *data.frame* into a list so that each of these objects represents one element of the list. Thus, lists can simply combine different objects into one instead of defining each as an individual variable. In this example, we use the variables created earlier, define a name for each and put them into a list:

```
l <- list(our_vector = v, our_matrix = m, our_data_frame = df)
```

11.5 Data indexing (subsetting)

While the table with all data is crucial for an overall analysis, in most cases only parts of it are required for an actual analysis or plot. For example, you may be interested solely in the first column or in extracting the 2nd to 10th value for a subsequent analysis. Such operations are called indexing (or subsetting). In R, square brackets are used for such operations:

```
# vector
v[3]  # extracts the 3rd value of the vector
v[3:5]  # extracts the 3rd to 5th value

# matrix
m[1, 2]  # extracts the value which is located in the first
row, second column
m[1, 1:2]  # all values in the first row and 1st to 2nd column
```

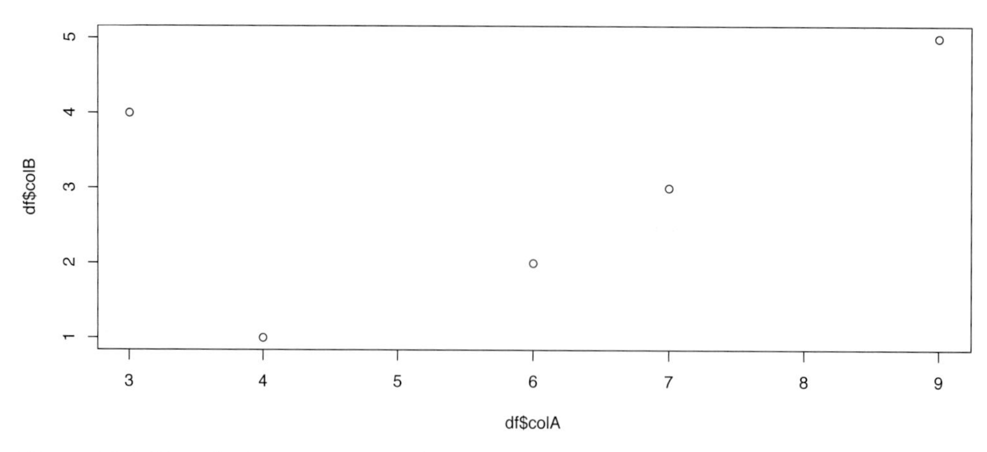

Figure 11.4 Plot of column A versus column B.

```
# data.frame
df[1, 2]  # same as for the matrix, first row, second column
position
df[1, 1:2]  # same as for the matrix, first row, 1st to 2nd
column
```

The overall approach for querying data in these three data classes is similar, but data frames offer a further option, by using the $ sign, to subset your data by column names:

```
df$colA
```

We can also use this subsetting approach to define which columns should be plotted (Figure 11.4):

```
plot(df$colA, df$colB)
```

The different options to index a data frame are shown in Figure 11.5. Different value ranges can be selected based on position or column name. A similar approach is needed when working with spatial raster or vector data and is introduced in the next chapter.

In certain cases, we may need to perform more complex queries, such as subsetting a column based on the values of a second column. For example, we may want to find all

df[, 1]
df$var1
df[2, c(2:4)]
df[c(1:4), 4]
df[1,1]
df[3,1]
df[3,]

	var1	var2	var3	var4
1	0.08	83	A	1
2	0.34	65	B	1
3	0.12	87	C	0
4	0.81	19	D	1

Figure 11.5 Indexing of single values, rows, columns or ranges for both rows and columns of a data frame.

field site names (colB) where the sample value (colA) is higher than 4. Performing such an operation on our data frame looks like this:

```
df$colB[df$colA > 4]

#> [1] b c e
#> Levels: a b c d e
```

Further Boolean operators can be used to query the data, such as equal (==), equal or greater than (>=) or unequal (`!=`). Additionally, different queries or conditions can be combined using 'or' (`|`) and 'and' (`&`).

11.6 Importing and exporting data

R offers specific functions that can read and load data from different file formats into a running R session. Common file formats across disciplines are supported by the core packages, whereas functions handling more specific file formats (e.g. spatial vector or raster data) are provided by specific packages, which require additional installations. Statistical data are often saved as tables, for example, in the comma-separated value (CSV) format. With the function `read.table()` you can read table files simply by providing the file path. The function `read.csv()` is specifically designed to read CSVs.

Have a look at the help of `read.table()` to make yourself familiar with the function and its options, for example, the arguments you can define. `read.table()` allows you to state how the values contained in your file are separated so that the function knows where to split the data into columns. Next, try these functions on a CSV file of your choice you have on your computer:

```
df <- read.table("path/to/mytable.csv", header = T)
df <- read.csv("path/to/mytable.csv")
```

Saving data from R to a table file, such as a CSV file, for example, after you have performed data manipulation, is as simple as reading it. Simply use `write.table()` or `write.csv()` to export your *data.frame*. Run the following functions to write your data back to a new file on your hard drive:

```
write.table(df, file = "path/to/mynewtable.csv")
write.csv(df, file = "path/to/mynewtable.csv")
```

The file is saved by defining the path with the export file name "path/to/filename.csv". If no path is added, it will be saved to your current working directory. Paths can be written as *relative paths* or *absolute paths*. A relative path is a path relative to your current location (run `getwd()` to check it), whereas the absolute path is the full path from your top directory. If you work in a specific working directory (e.g. a project folder on your hard drive), you can tell R where this working directory is using the function `setwd()`:

```
setwd("C:/path/to/your/working/directory")
```

If you have defined a working directory, you can use relative paths for every function that needs a file path, meaning you can skip the path to your working directory and do not have to (re)write it repeatedly. Moreover, the advantage of *relative paths* is that they will work for every user, once the user has defined where, on the device, the working directory is located.

TASK

Enter the relative and absolute path to save a file and then search for it in your file system. The relative path is the path viewed from your working directory, omitting the whole path above the working directory. Absolute paths represent the full path from the drive's root directory to the actual file or folder.

11.7 Functions

All the commands we have used so far in the chapter are *functions*. Functions are the backbone of R and allow you to perform a certain action to an object in your R environment, for example, saving a table to your hard drive. In general, functions combine all the code you would have to write out each time you wanted to perform the specific task the function represents. For example, the function `write.table()` contains all the code necessary to write a *data.frame* to a file on disk. Thus, instead of having to create the code for common functions in R by yourself, you can use available functions that will do the task for you.

Building your own functions in R is also very easy. Just imagine you have a basic task you need to apply very often, for example, calculating the normalized difference vegetation index (NDVI). Instead of physically writing the formula of the NDVI to your script each time you calculate the NDVI, you can also mainstream the calculation by writing the formula once inside a function and then calling this function each time you want to calculate the NDVI.

Like the formula example we outlined earlier, you can define a function to take different arguments. Arguments are inputs to the function, for example, the RED and near-infrared bands of different data sets. The input arguments are then used by the function code regardless of which data set you use.

First, create a simple function by executing the following:

```
your_function <- function(x, y) {
   z <- x + y
   return(z)
}
```

Note that you can name the variables anything you want, x and y or A and B, or use more meaningful names. Can you guess from the code what this function does? It simply takes two inputs (arguments x and y), adds them and saves the result to a variable named z, which is then returned by the function. Give it a try and call your own function with different values:

```
your_function(x = 5, y = 10)
```

Of course, this example function is not very useful. A function becomes particularly useful if the operation inside the function is more complex than just simple addition. Writing functions can make it easier to keep an overview of where, in your script, each calculation is performed and can reduce the potential for errors, since the code to be executed is written only once instead of several times. When writing a script, always try to reduce as much redundancy as possible to avoid errors.

11.8 Loops

Sometimes, you may want to perform the exact same action repeatedly, for example, calculating the NDVI for a long list of raster files or values. First, we create an example list for which we can demonstrate the advantage of a loop by producing 30 random numbers between 1 and 100 and putting them into a list:

```
my_list <- as.list(sample(1:100, 30))
print(my_list)
```

When you print `my_list`, the content is displayed in the console. If we now want to perform a task for each element in the list, for example, multiply each value by 10, we could write the code for each element separately:

```
result1 <- my_list[[1]] * 10
result2 <- my_list[[2]] * 10
result3 <- my_list[[3]] * 10
result4 <- my_list[[4]] * 10
result5 <- my_list[[5]] * 10
```

However, this can be a time-consuming job, especially if we are dealing with long lists. Thus, instead of writing 30 lines of code like this, we can simply loop through the list and apply the task subsequently to each element in the list:

```
results <- list()
for (i in 1:length(my_list)) {
  results[[i]] <- my_list[[i]] * 10
}
```

In this example, `list()` creates an empty list that will store all the results created by the loop. To loop over the list, a `for` loop is applied with `i` being an index that starts at 1 and ends at the length of `my_list` (in our example, it is automatically queried by `length()`, which is 30). This means that the `for` loop will iterate 30 times through the code within the loop, each time changing index `i` to the next higher number and thus running the calculation (in this example, × 10) for each element in the list. The result of each calculation is then saved as a separate element to a new list, the `results` object.

Apart from '`for`' loops, there are also `while` loops. Instead of iterating through a predefined number of items, `while` loops are conditional, which means they repeat an action indefinitely until a certain logical condition is fulfilled. There are also other ways of building a loop in R. Some are shown in the following chapters. If you are interested in learning more about iteration techniques and conditions in R, we recommend reading *R for Beginners* (https://cran.r-project.org/doc/contrib/Paradis-rdebuts_en.pdf).

11.9 Scripts

In the long run, we highly recommend that you to write a separate script for each analysis and add as many explanations or comments to your code as necessary. For example, it is a good habit to add comments about the intention of each step (not what the actual function does), even if you think it is obvious. You will be thankful if, several months later, you need go back to your script and try to understand what you were doing back then. Also, it makes your code transparent and understandable to other people.

Thus, a script is in fact just a series of functions saved in one file. Comments not executed by R are indicated by # and can be used to explain the code in natural language. An example of a good script with an appropriate code-to-comments ratio would look something like this:

```
# A short description of your script
# #############
# Purpose: This is a script to do analysis xy. Data sets will
be ...
# Author: your name
# Date: month, year
# R version and packages: x.xx
# ##############
# ##############
# Import format: csv, tab-delimited
# Output format: plot
# ##############

# Import the csv with x and y coordinates
data_in <- read.table("/path/to/mytrack.csv")

# check first entries
head(data_in)

# overview statistics
summary(data_in)

# plot data
plot(data_in)
```

11.10 Expanding functionality

R starts with a minimal set of functions available through default packages. To expand the functionality of R, you need to install additional packages using the function `install.packages()`. When prompted to pick a repository, simply select one in your vicinity. After the installation, you must load the required package at the beginning of each session using the `library()` function, otherwise you cannot use its functionality. To keep R running quickly, we suggest you only load those packages you really need for the current session.

Install and activate the 'ggplot2' package or any other package you might need. Installation only needs to be done once, activation (`library()`) must be executed always at the beginning of a new script.

By default, packages are installed from the Comprehensive R Archive Network (https://cran.r-project.org), a global network of servers to which developers can submit their packages. If you encounter an error or warning stating that a package is not available, check whether:

- You have misspelled the name of the package.
- You have the most recent version of R installed.
- The package is available on the Comprehensive R Archive Network or must be installed from another source. (Usually, installation guidance is provided on the respective package Web page.)

If a function does not work and/or R tells you that it cannot be found, you may simply have forgotten to load the package providing this function. In this case, find out which package the function you want to use belongs to and load the package with the `library()` function.

11.11 Bugs, problems and challenges

We have provided you with working examples. However, the behaviours of different operating systems, changes in software functionality or command syntax are possible. Thus, you will have to learn how to interpret error messages. Most importantly, please read error messages carefully; usually, errors are resolved within minutes that way.

Generally, error messages are quite logical, for example, consider this one:

```
x <- raster("your_file.tif")

#> Error in eval(expr, envir, enclos): Error in raster("your_
file.tif") : could not find function "raster"
```

This error message tells you that the function raster cannot be found; hence, you simply forgot to load the related library. Execute `library(raster)` and your function will work.

Another error message we have seen quite often, and which confuses beginners, is:

```
library(raster)
library(rgdal)

x <- raster("your_file.tif")

#> Error in eval(expr, envir, enclos): Cannot create a
RasterLayer object from this file. (file does not exist)
```

This message indicates that your file does not exist where R thinks it should be. However, if you add the correct path or name, the function should work without problems. As you can see, reading the error message gives you a pretty good idea as to what the problem might be.

Problems are very useful because as you learn how to deal with them, each error message is a *lesson learnt*. If you do not understand the error or warning message you are faced with, copy and paste it to your favourite Web search engine. It is very likely someone else has encountered the same or a similar problem and a solution to the problem has been posted online.

11.12 Notation

When mentioning some R functions in the chapters, we might refer to their full name to make it easier for you to see which package the function comes from. Instead of writing: 'use `myFunction()` from package *myPackage*', we will write `myPackage::myFunction()`, which is a valid expression in R. Usually, R finds the correct function from the correct package automatically. However, in rare cases, you can run into trouble when different packages define a function using the same name. For example, the `spTransform()` function can be found in the packages *move*, *rgdal* or *sp*, and the `aggregate()` function can be found in the *stats* and *raster* packages. In such a case, it is a good idea to explicitly define which function from which package you mean:

```
myAggRaster <- raster::aggregate(myRaster, 2)
```

The next chapters provide details about the spatial realm of R coding, where the functions learnt in this chapter can be executed on spatial data. Analysis, and spatial and non-spatial mapping, are also be covered.

11.13 Summary and further reading

In this chapter, you learnt what a CLI is and how it differs from a GUI. You were introduced to the programming language R, its classes and its basic syntax, and learnt your first functions. We explained how to read and write data, what functions and loops are and how to arrange a script. Several books on learning R exist and we mentioned a few. If you would like to learn more about the power of R (spatial and non-spatial) we recommend reading *Getting Started with R* by Andrew Beckerman, Dylan Childs and Owen Petchey, *R for Data Science* by Garrett Grolemund and Hadley Wickham, and *The Art of R Programming: A Tour of Statistical Software Design* by Norman Matloff. We also recommend that you search for new books on R programming since the language is evolving fast; with thousands of available packages, some even peer-reviewed, there is always something new to learn about R.

12. Getting started with spatial coding

Dealing with spatial data in R is as simple as dealing with non-spatial data. Whereas the core packages of R (those packages installed right from the beginning that are required to make R work) are not specifically designed to handle spatial data, the R community has been developing a variety of powerful and dedicated packages for spatial data handling and analysis over the years. Some of these packages have received growing acceptance throughout the spatial community, making them the standard for basic functionalities, such as reading and writing vector or raster data and performing basic analysis. Other spatial packages offer very specific tools that may be interesting in some cases but are not required for day-to-day tasks. The major packages we rely on to turn R into fully fledged spatial data analysis software are:

- *raster*, which provides representations for spatial raster data and offers tools to manipulate and analyse it.
- *rgdal*, which provides tools to read and write spatial data in a variety of formats and for projection/transformation operations.
- *RStoolbox*, which provides tools specifically designed for remote sensing data analysis, such as calculating indices or creating classifications.
- *sf*, which provides representations for spatial vector data (defined as simple features) and functions for spatial manipulations, such as geometric subsetting and intersecting.
- *getSpatialData*, which provides tools to query, preview and download satellite sensor imagery and products.

To get a more extensive overview over spatial packages, check the CRAN Task View: Analysis of Spatial Data Web page for packages dealing with spatial data (https://cran.r-project.org/web/views/Spatial.html). There is also a specialized help list at R-sig-Geo, the R Special Interest Group on Using Geographical Data and Mapping, where you can post questions concerning spatial data analysis in R (https://stat.ethz.ch/mailman/listinfo/r-sig-geo).

12.1 Spatial data in R

Several packages in R provide spatial data handling functionalities, such as the *raster* or *sf* packages. These packages are installed once and then activated each time a new analysis

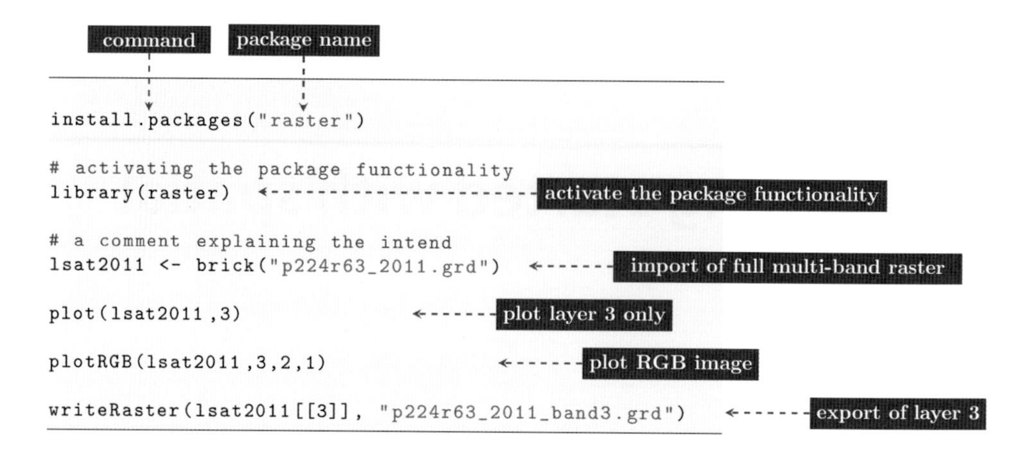

Figure 12.1 General functions for working with spatial raster data, including the installation and activation of necessary R packages.

is required (Figure 12.1). Please install all the packages listed at the start of this chapter using `install.packages()` and load them using `library()`.

A good start into the spatial realm of R coding is feasible with some simple functions provided by the *raster* package. We can easily download country borders (Figure 12.2). In this example, we download the borders of Germany as vector data, save them to a variable called '`germany`' and then plot them:

Figure 12.2 Plot showing the boundaries of Germany.

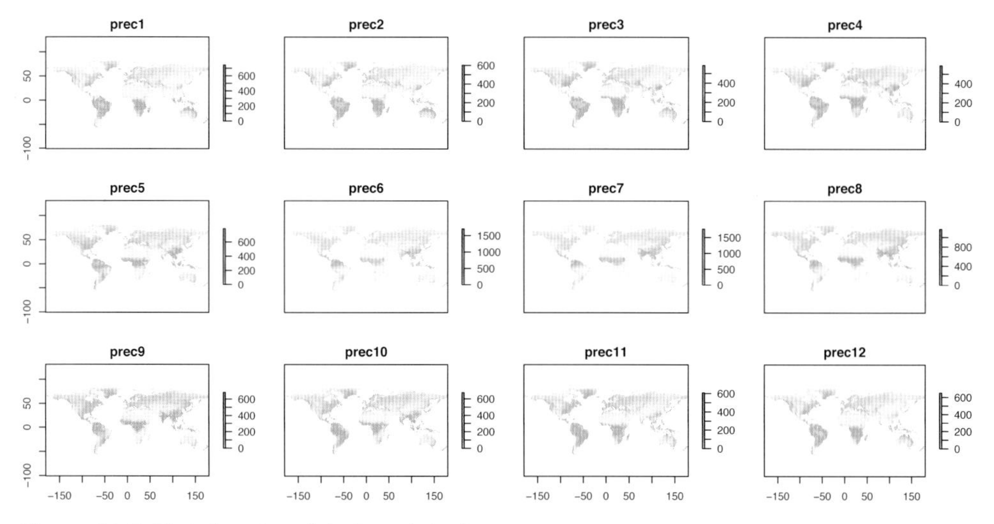

Figure 12.3 Plot showing global precipitation.

```
library(raster)
germany <- getData("GADM", country = "DEU", level = 1)
plot(germany)
```

The *raster* package also allows us to download rasterized climate data without the need for a time-consuming manual search on the Web. In this example, we download precipitation data for the whole world and then plot them (Figure 12.3):

```
precip <- getData("worldclim", var = "prec", res = 2.5)
plot(precip)
```

By simply calling the name of a variable, a summary of its contents is displayed. In this example, the variable 'germany' is a polygon vector, called *SpatialPolygonsDataFrame*, with 10 variables (attributes) and 16 features (single polygons):

```
germany

#> class       : SpatialPolygonsDataFrame
#> features    : 16
#> extent      : 5.866251, 15.04181, 47.27012, 55.05653 (xmin,
xmax, ymin, ymax)
#> crs         : +proj=longlat +datum=WGS84 +no_defs
+ellps=WGS84 +towgs84=0,0,0
#> variables   : 10
#> names       : GID_0, NAME_0, GID_1, NAME_1, VARNAME_1, NL_
NAME_1, TYPE_1, ENGTYPE_1, CC_1, HASC_1
#> min values  : DEU, Germany, DEU.1_1, Baden-Württemberg,
Bavaria, NA, Freie Hansestadt, State, 01, DE.BE
#> max values  : DEU, Germany, DEU.9_1, Thüringen, Thuringia,
NA, Land, State, 16, DE.TH
```

The variable `precip`, on the other hand, is a raster object called *RasterStack* because it consists of several raster layers of the same spatial extent and resolution that are stacked together:

```
precip

#> class       : RasterStack
#> dimensions  : 3600, 8640, 31104000, 12 (nrow, ncol, ncell,
nlayers)
#> resolution  : 0.04166667, 0.04166667 (x, y)
#> extent : -180, 180, -60, 90 (xmin, xmax, ymin, ymax)
#> crs         : +proj=longlat +datum=WGS84 +ellps=WGS84
+towgs84=0,0,0
#> names       : prec1, prec2, prec3, prec4, prec5, prec6,
prec7, prec8, prec9, prec10, prec11, prec12
#> min values  : 0, 0, 0, 0, 0, 0, 0, 0, 0, 0, 0, 0
#> max values  : 916, 790, 818, 871, 1118, 2386, 2437, 1763,
1121, 1132, 1005, 975
```

Apart from the variable class, additional information that is relevant for us is also displayed, namely the coordinate reference system (CRS) and the names of the layers with their value ranges. You might wonder how these spatial objects relate to non-spatial objects, such as a data frame or matrix. In fact, they are very similar except that geographical information is attached, putting the values displayed into a geographical context. Using `germany@data`, we can easily extract the values from a spatial object.

This separates the geographical information from the pure values, which are represented by a table. The various columns provide a range of information stored within our vector object. In our example, a single row provides data related to one specific state (*NAME_1*) within the country Germany (*NAME_0*). Depending on the data set, information such as population or divisions that are further subdivided spatially (*municipalities*) is also provided.

Similarly, we can extract all values of a *RasterStack*:

```
precip[]

#>       prec1 prec2 prec3 prec4 prec5 prec6 prec7 prec8 prec9
#> [1,]      7     8     9     7     6    12    15    22    20
#> [2,]      7     8     9     7     6    12    15    22    20
#> [3,]      7     8     9     7     6    12    15    22    20
#> [4,]      7     8     9     7     6    12    15    22    20
#> [5,]      7     8     9     7     6    12    15    22    20
#> [6,]      7     8     9     7     6    12    15    22    20
#> [7,]      7     8     9     7     6    12    15    22    20
#> [8,]      7     8     9     7     6    12    15    22    20
```

The data set `precip` consists of several columns with a very long list of values. The columns relate to the months; the single values relate to each pixel. The actual content of

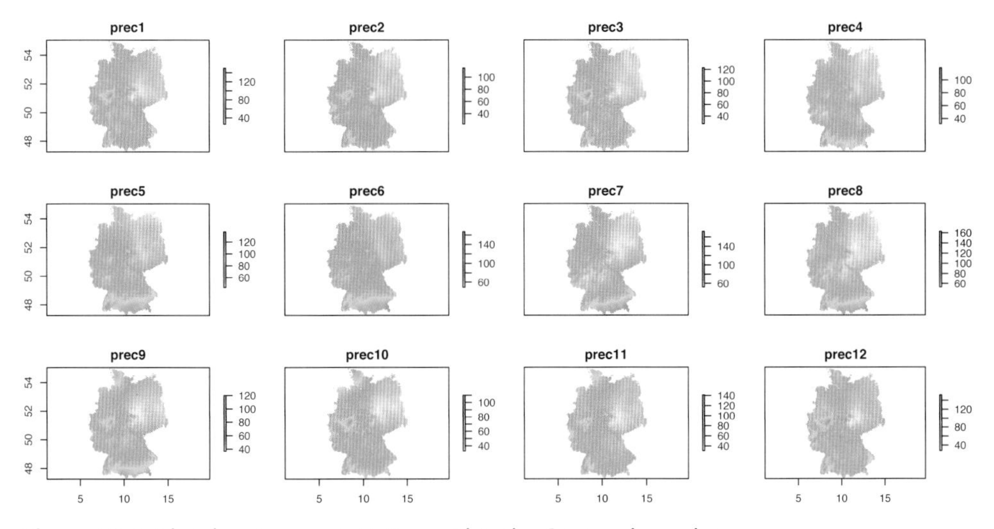

Figure 12.4 Plot showing precipitation within the German boundary.

each data set is provided within the respective manual, where the row names and values are explained.

The environmental data set covers the whole globe, whereas we only require the data for our study area; therefore, we reduce the data set to the extent of Germany and mask out all values outside the shape of Germany (Figure 12.4). This is done using similar functions to those already executed in QGIS (Figure 12.5) but with slightly different names. First, we crop or cut our precipitation raster to the extent of the vector object '`germany`'. This results in a rectangular precipitation raster extent:

```
precip_crop <- crop(precip, germany)
```

To remove all values outside the boundary of Germany, we mask the cropped precipitation raster to the boundary of our vector:

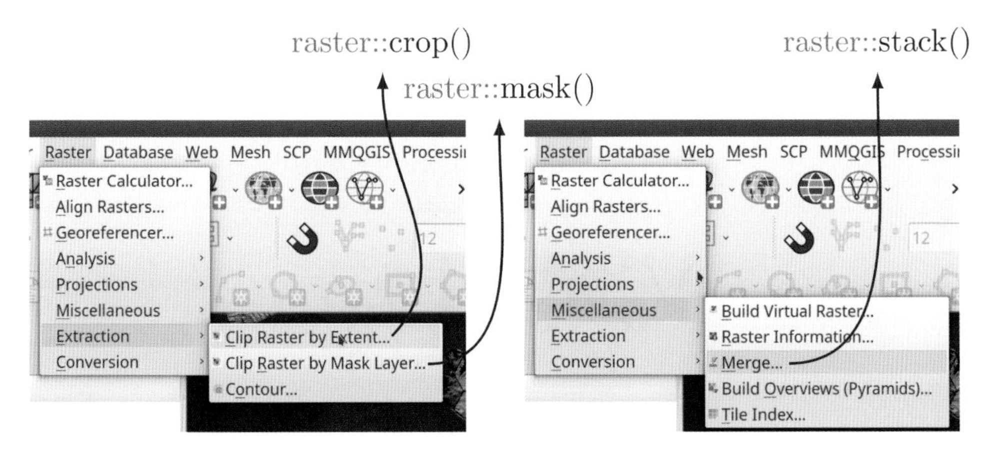

Figure 12.5 Spatial data can be masked and clipped in R in similar fashion as using the functionalities provided by QGIS.

```
precip_crop_mask <- mask(precip_crop, germany)
```

Now we have the precipitation values within the boundary of Germany:

```
plot(precip_crop_mask)
```

Doing such operations in R and creating new spatial objects based on a certain function is similar to how you would do this using the QGIS interface. However, using R functions like the ones used so far makes it easier to store the operation chain of a spatial analysis or share it with colleagues. Moreover, it can be easily executed again combined with other data sets. For example, you could do the same task using another country simply by changing the country abbreviation or downloading a different environmental parameter.

> **TASK**
>
> - Use other countries (see the ISO code in the manual).
> - Use other environmental data sets provided by `getData()`.
> - Swap mask and crop and check the difference in the output.

12.2 Importing and exporting data

So far, we have imported data from the Web. However, usually the spatial files you require for your project are on your local hard drive. All data must be imported into R to process them or exported from R to store or use them elsewhere. Creating or modifying a spatial object in R will not automatically change the original file on your hard drive. In addition, the *global environment* containing the results of an analysis are usually deleted when a session is closed. Everything that has not been saved to the hard drive is lost. In the following sections, the data we have used before will be imported and processed within R. The examples are split into vector and raster formats because of the different packages and functions involved.

12.2.1 Vector data

Vector data can be imported using various functions depending on the package used and the format of the data. For example, a text file with coordinates and point values can

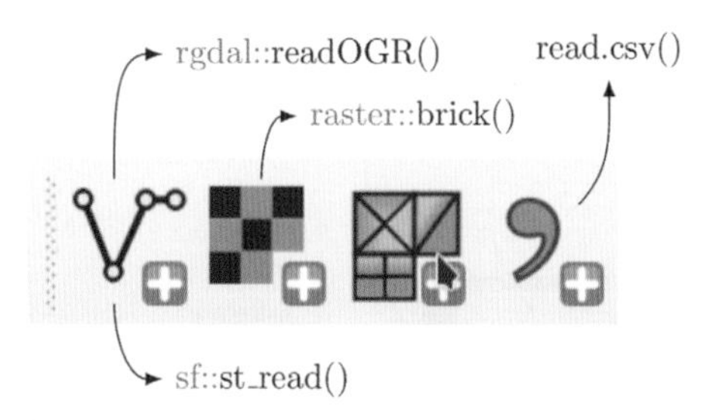

Figure 12.6 Importing spatial data in R is like using the load functions in QGIS. `readOGR()` can be used to import spatial vector objects, while `read.csv()` can be used to import text files.

be imported with the `read.csv()` function and converted to a spatial object like the corresponding QGIS function. A direct import of a spatial vector object can be done with `st_read()` or `readOGR()` (Figure 12.6).

12.2.2 Non-spatial vector data

A common task is to use existing coordinates within a data frame and convert them into a spatial object. In Chapter 5, we performed the equivalent operation in QGIS using a text file with coordinates and attributes (*coordinates_with_attributes.csv*). In QGIS, the text file is converted straightaway into a spatial object. In R, we first need to import the text file:

```
df <- read.csv("coordinates_with_attributes.csv")  #from
Chapter 5
```

We can look at this imported data frame `df` by calling its name. Alternatively, you can call `summary(df)` to see a summary of the object or call `View(df)` to see the object's full content. This object consists of two columns with coordinates plus three columns with information about environmental conditions and field site names. To preserve the original object, for example, for other analysis, we copy the data to another object using a name that indicates a spatial point data frame (spdf) structure, because we will convert this data frame into a spatial object in the next step:

```
spdf_df <- df
```

Changing the non-spatial data frame to a spatial data frame requires the *sp* package and is done by defining the columns to be used as coordinates. In this case, they are:

```
library(sp)
coordinates(spdf_df) <- ~xcoord + ycoord
```

You should always check the names of the coordinate columns beforehand. The coordinate columns can also be named *lat* and *long* or just *x* and *y*, so the function will have to be adapted accordingly. Calling the summary information of `spdf_df` shows us that it is now a spatial object:

```
summary(spdf_df)

#> Object of class SpatialPointsDataFrame
#> Coordinates:
#>            min       max
#> xcoord  609752  621444.7
#> ycoord 5525178 5534449.3
#> Is projected: NA
#> proj4string : [NA]
#> Number of points: 6
#> Data attributes:
#>         id         measurement1   measurement2       name
#>  Min.   :0.00   Min.   :0.0   Min.   : 3.00   plotA:1
#>  1st Qu.:1.25   1st Qu.:0.0   1st Qu.: 9.75   plotB:1
```

```
#>  Median :2.50   Median :0.5   Median :16.50   plotC:1
#>  Mean   :2.50   Mean   :0.5   Mean   :16.50   plotD:1
#>  3rd Qu.:3.75   3rd Qu.:1.0   3rd Qu.:22.50   plotE:1
#>  Max.   :5.00   Max.   :1.0   Max.   :31.00   plotF:1
```

The points of this spatial data frame can be plotted directly and will no longer be displayed as a scatterplot, rather as a spatial map. However, the CRS is not yet stored within this object; therefore, all subsequent operations with projected objects will not work. In this case, the CRS must be known and assigned by the user. Because we created this text file, we also know the projection. For any future work involving coordinates within text files, you must always record the projection used! Assigning the CRS is done either by using the actual name of the projection or through the European Petroleum Survey Group code:

```
proj4string(spdf_df) <- CRS("+init=epsg:32632")
```

Now, we have a spatial point data frame with a projection. If you call the name again, you can see that the CRS has been supplemented by the projection information. Further related data can also be added to our newly generated spatial vector object. In this example, the same data is used but any other corresponding data frames could also be added to the spatial vector object:

```
spdf_df@data <- df
```

Calling this object shows that the structure is spatial, including the projection and different attributes. You should always check if the data and location have been converted correctly. This object can now be exported as a spatial object for later use choosing a variety of formats. In this example, we use the GeoPackage format, an open format that stores all information in one file, and export it using the `writeOGR()` function in the *rgdal* package:

```
library(rgdal)
writeOGR(spdf_df, dsn = "coordinates_with_attributes.gpkg",
layer = "spdf_df",
    driver = "GPKG")
```

Alternatively, you can export this object using the *sf* package, the updated successor of the *sp* package. While *sp* is still used, especially since it is highly compatible with the *raster* package, the *sf* package is gradually replacing the functionalities of the *sp* package. Therefore, while we introduce both packages, we focus on the *sf* package in the following sections. We convert our *sp* object into an *sf* object and then use the *sf* function `st_write()` to write the object to a GeoPackage file:

```
library(sf)
spdf_df_sf = st_as_sf(spdf_df, wkt = "geom")
st_write(spdf_df_sf, "coordinates_with_attributes.gpkg")
```

Once this is done, you should see the spatial file on your hard drive. As mentioned previously, the file is saved to your current working directory. (You can check this with `getwd()` or you can add a full path within the export function that specifies where to save the object (“`data/vector_data/coordinates_with_attributes.gpkg`”).)

12.2.3 Spatial vector data

The previously exported file and a wide range of spatial vector or raster data products can be imported directly without any coordinates or assigned projection, since spatial information is stored within the corresponding files. This can be done using the respective spatial functions and plotted using the *rgdal* or *sf* functions:

```
# using rgdal
field_sites <- readOGR("coordinates_with_attributes.gpkg")
plot(field_sites)

# using sf
field_sites <- st_read("coordinates_with_attributes.gpkg")
plot(st_geometry(field_sites))
```

After the spatial objects have been imported, they can be used for further changes and analysis.

12.2.4 Raster data

Like spatial vectors, we can also import raster files. These are usually provided as GeoTIFF or GRD files, which store the projection, pixel size and coordinates. In this example, we use `brick()` to import our multilayer Sentinel-2 data set to a variable named `s2`:

```
s2 <- brick("T32UPA_A015792_20180701T102404_STACK_crop.tif")
```

Calling the name of this object provides details about its dimension, spatial resolution, projection and included bands. In this example, we have four bands included in our brick with a 10-m resolution and the *UTM WGS84 zone 32* projection.

Raster *bricks* are stacks that originate from a single multilayer file, as in this example. Single-band raster files can be imported using `raster()`. Raster objects that have matching grids, extents and projections can be combined into a stack, like a brick, using `stack()`. In this example, we import single-layer rasters and then stack them; alternatively, you can do the same by creating the stack directly from the single-layer files:

```
s2_b2 <- raster("T32UPA_A015792_20180701T102404_B02_crop.tif")
s2_b3 <- raster("T32UPA_A015792_20180701T102404_..._crop.tif")
...
s2 <- stack(s2_b2, s2_b3)

# or instead do it directly:
s2 <- stack("T32UPA_A015792_20180701T102404_B02_crop.tif",
"T32UPA_A015792_20180701T102404_..._crop.tif")
```

The resulting spatial raster object should always be checked for consistency (projection, extent, etc.). We also recommend plotting it with another spatial object to check if all data match spatially. In this example, we plot the first layer of our Sentinel-2 data set and overlay (`add=TRUE`) it with the spatial points imported previously (Figure 12.7):

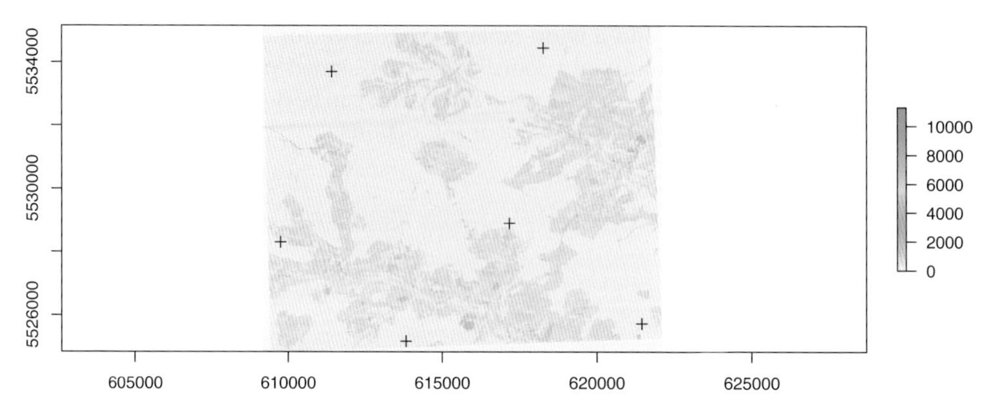

Figure 12.7 Band 1 of our Sentinel-2 raster stack with the superimposed point locations.

```
plot(s2, 1)
plot(spdf_df, add = TRUE)
```

Apart from the coordinates of the study area and the two spatial objects, the map also shows the legend of the raster data set. In Chapter 15, we introduce better ways to display spatial data using the *ggplot2* package.

12.3 Modifying spatial data

As already mentioned, various spatial operations can be applied to spatial data, such as subsetting, reprojecting or buffering. We have already encountered such operations in QGIS; matching functions are available in R.

12.3.1 Reprojecting vector data

To ease the application of spatial vector objects within the *sf* domain, we can either convert them in each function from *sp* to *sf* or convert the object once to *sf* and work with it:

```
library(sf)
sf_df <- st_as_sf(spdf_df)
```

Looking at this new object shows that all information is kept, but that the object has been transformed from a *SpatialPointsDataFrame* to a *simple features* data frame.

As in QGIS, there are dedicated tools to reproject vector and raster objects (Figure 12.8). In this example, we use `st_transform()` to reproject the vector object:

```
sf_df.reproj <- st_transform(sf_df, st_crs("+proj=latlong
+datum=WGS84"))
```

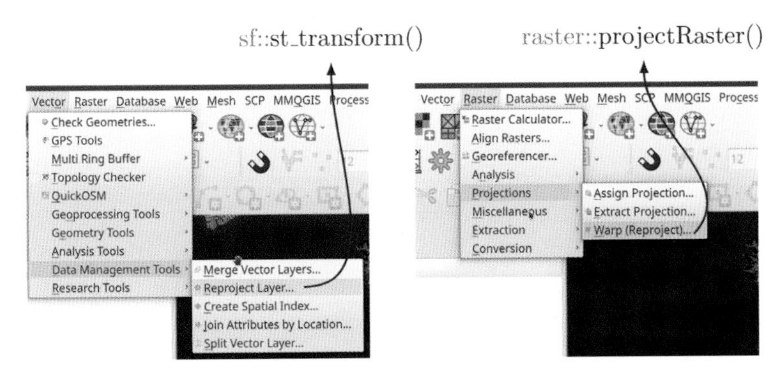

Figure 12.8 QGIS interface and corresponding R functions for reprojecting raster and vector data.

12.3.2 Geometry operations

Other operations, such as buffering, can also be achieved in R using functions that correspond to the ones we introduced in QGIS (Figure 12.9). For example, adding a 5,000-m buffer in *sf* is done by:

```
sf_df_buffer5000 <- st_buffer(sf_df, dist = 5000)
```

Before buffering, our vector object was represented by points and looked like this (Figure 12.10):

```
plot(st_geometry(sf_df))
```

Through buffering, it has been transformed to a vector object represented by polygons (Figure 12.11):

```
plot(st_geometry(sf_df_buffer5000))
```

Similarly, geometry operations such as unions, intersections or differences can be performed using the functions `st_union()`, `st_intersection()` or `st_difference()` (Figure 12.12):

```
geometry_union <- st_union(sf_df_buffer5000)
plot(st_geometry(geometry_union))
```

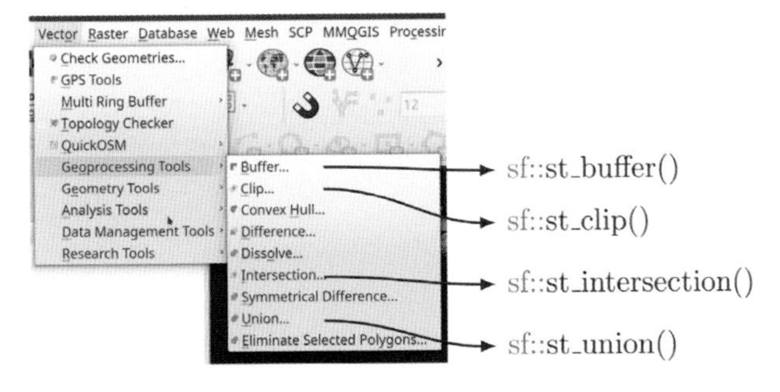

Figure 12.9 Vector operations in QGIS and their equivalent in R using the sf package.

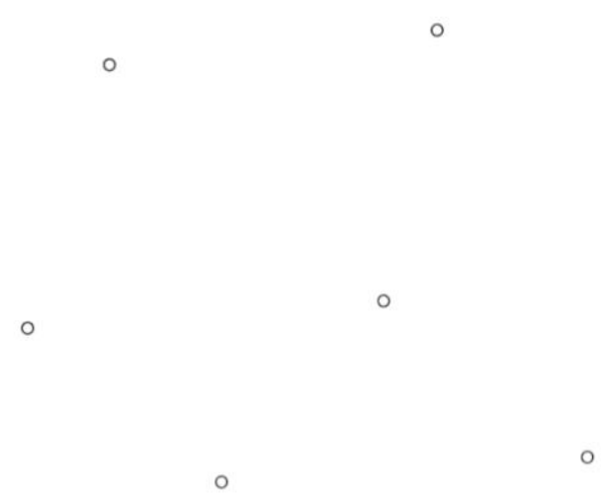

Figure 12.10 Points before buffering.

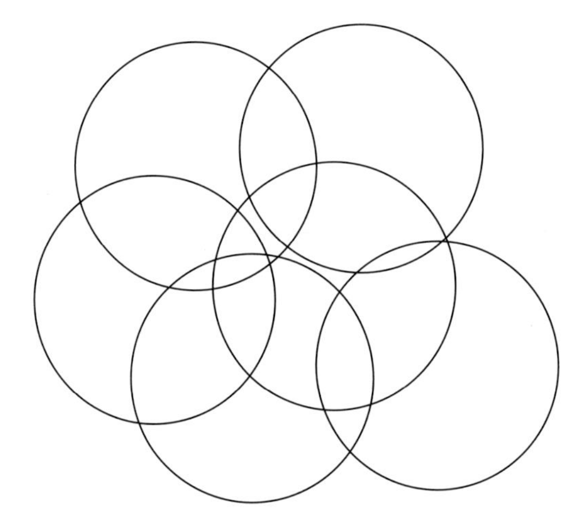

Figure 12.11 Resulting 5,000-m buffer using the `st_buffer` function.

Figure 12.12 Output of the `st_union` function on the buffer polygons.

TASK

Run `st_intersection()` and `st_difference()` on the vector data, compare their outputs and see how they differ. Find other related functions in the manual pages.

12.3.3 Querying and subsetting vector data

Very often, not only the whole vector is of interest but also parts of it, such as provinces or single cities. Any spatial object can be queried or subsetted based on values in the data table associated with it. For example, we can extract just the polygon for the state of Bavaria, which is listed in the *NAME_1* column. Please note that in this column, the German name (Bayern) is stored and thus must be used:

```
germany_by <- germany[germany$NAME_1 == "Bayern", ]
```

The square parentheses are a function that tells R to extract only the polygon within the `germany` vector, which lists the character "`Bayern`" in the *NAME_1* column. The comma at the end is not a mistake, rather it is needed because we are subsetting by row.

Before subsetting, our vector object represented the whole of Germany (Figure 12.13):

```
plot(germany)
```

The object resulting from subsetting by name, which represents the boundaries of Bavaria, looks like this (Figure 12.14):

```
plot(germany_by)
```

This can be done for any value or character within our vector. For more complex

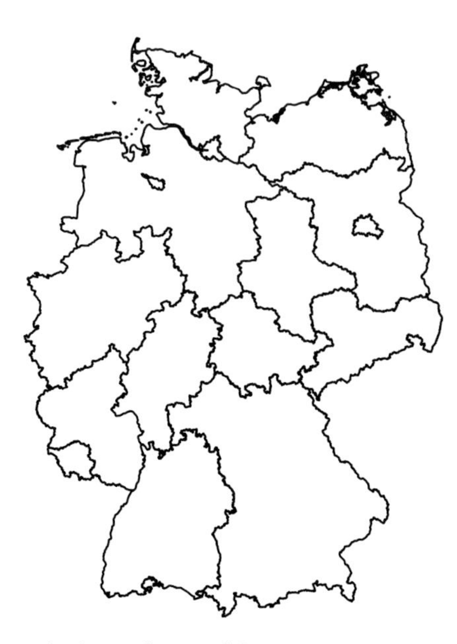

Figure 12.13 Plot showing the boundaries of Germany.

Figure 12.14 Plot showing the boundaries of Bavaria.

operations, it is also possible to combine operations or query all polygons with values higher or lower than a certain threshold.

Like the QGIS Field Calculator, we can query and subset our spatial objects by the values within the table:

```
sf_df_sel <- sf_df[sf_df$measurement2 > 20, ]
```

All locations where the *measurement2* column has a value >20 have been selected.

12.3.4 Raster data

Raster data can also be modified in R, for example, by multiplying two raster layers:

```
s2.1x2 <- s2[[1]] * s2[[2]]
```

The resulting raster object can be plotted with the `plot()` function. Reprojecting a raster object is done using `projectRaster()` or changing its resolution with `resample()`. Additional raster functions can be found in the *raster* package manual. Examples on how to modify raster data in R are provided in Chapter 13.

12.4 Downloading spatial data from within R

Many spatial raster and vector data sets can be accessed and downloaded from online sources directly within R. The *raster* package contains the function `getData()`, which can download basic, temporally static raster and vector data sets for specific regions, for example, climatic data, elevation data or country borders. Check the help of `getData()` to find out which data sets are available. The following example demonstrates how to download a Shuttle Radar Topography Mission (SRTM) digital elevation model tile covering a geographical location defined by longitude and latitude:

```
library(raster)
dem_tile <- getData("SRTM", lon = 10.55, lat = 49.92)
```

TASK

Practise with `getData()` and its arguments to download different data. Read the `getData()` manual by calling `?getData` to see how to use the function.

Data downloaded using the *raster* package are temporally static, meaning they represent a single point in time instead of change over time. Data change with time, such as satellite imagery acquired within a certain temporal interval, can be downloaded using the *getSpatialData* package.

With *getSpatialData*, you can query, preview, filter and download satellite sensor data as demonstrated using the *SCP* QGIS plug-in in Chapter 2, but without relying on a graphical user interface. The following is a short example on how to query both Sentinel-2 and Landsat-8 data for the area of interest (AOI) used in this book.

First, install and load *getSpatialData* and *sf*, if not already done:

```
library(getSpatialData)
```

Then, define an AOI. You can load and use the AOI we have been using in this book:

```
aoi <- st_read("Chapter_02/AOI/aoi.shp")
set_aoi(aoi$geometry)
```

The AOI is now saved and will be used for every data query you perform during the current R session. Alternatively, you can draw an AOI of your choice on a map by calling `set_aoi()` without defining a path:

```
set_aoi()
```

Before running a query, you must log in to the services you want to query. Use the same login credentials you used to search and download data with QGIS in Chapter 2:

```
# login to Copernicus Hub (you will be asked for your
password)
login_CopHub(username = "your_username")
# if you do not have a Copernicus Hub account, create one
here:
# https://scihub.copernicus.eu/dhus/#/self-registration

# login to USGS ERS
login_USGS(username = "your_username")
# if you do not have a USGS ERS account, create one here:
# https://ers.cr.usgs.gov/register/

#> Login successful. ESA Copernicus credentials have been
saved for the current session.

#> Login successful. USGS ERS credentials have been saved for
the current session.
```

Additionally, *getSpatialData* needs to know where it can save the data you want to download. You can define an archive directory to which the package will download all files ordered by sensor using `set_archive()`:

```
set_archive("path/to/a/folder")
```

Now, *getSpatialData* is ready to search for data. To get an overview of which data products are available, run `get_products()`:

```
# get an overview of all available product names
get_products()

#>  [1] "Sentinel-1"       "Sentinel-2"      "Sentinel-3"
#>  [4] "Sentinel-5P"      "LANDSAT_ETM_C1" "LANDSAT_MSS_C1"
#>  [7] "LANDSAT_TM_C1"    "LANDSAT_8_C1"    "MODIS_MCD43D28_V6"
#> [10] "MODIS_MYD21A1N_V6"
```

Pick those products for which you want to run a query for your AOI and a certain time range. In this example, we have picked *Sentinel-2*; however, you can also query multiple product names at once, for example, *Sentinel-2* and *LANDSAT_8_C1* (Landsat 8, collection 1).

```
# query for Sentinel-2 records
records <- get_records(time_range = c("2019-07-01", "2019-07-
10"), name = "Sentinel-2")
#> Searching records for product name 'Sentinel-2'...
```

The variable *records* is a data frame containing all records found for the selected products, time range and AOI. Have a look at the resulting table:

```
View(records)

#> Simple feature collection with 16 features and 2 fields
#> geometry type:  MULTIPOLYGON
#> dimension:      XY
#> bbox:           xmin: 8.999718 ymin: 49.52842 xmax:
11.95939 ymax: 50.55229
#> epsg (SRID):    4326
#> proj4string:    +proj=longlat +datum=WGS84 +no_defs
#> First 10 features:
#>    product_group    product                       footprint
#> 1       Sentinel Sentinel-2 MULTIPOLYGON (((10.47044 49...
#> 2       Sentinel Sentinel-2 MULTIPOLYGON (((10.47032 49...
#> 3       Sentinel Sentinel-2 MULTIPOLYGON (((10.47044 49...
#> 4       Sentinel Sentinel-2 MULTIPOLYGON (((10.47032 49...
#> 5       Sentinel Sentinel-2 MULTIPOLYGON (((10.51786 49...
#> 6       Sentinel Sentinel-2 MULTIPOLYGON (((11.89936 49...
#> 7       Sentinel Sentinel-2 MULTIPOLYGON (((11.89936 49...
```

```
#> 8        Sentinel Sentinel-2 MULTIPOLYGON (((10.51786 49...
#> 9        Sentinel Sentinel-2 MULTIPOLYGON (((10.47011 49...
#> 10       Sentinel Sentinel-2 MULTIPOLYGON (((10.46999 49...
```

Check the spatial footprints of all records to see if they match the area you are investigating (Figure 12.15):

```
view_records(records)
```

If you want to preview records before downloading them, you can tell *getSpatialData* to collect preview images for all or some of the records you just queried. Preview images are small, georeferenced RGB images with a much lower spatial resolution than the actual data, which allow you to visually inspect a data set before downloading it. With `view_previews()`, you can display the preview images on a map (Figure 12.16). In this example, we want to view records 5–8.

```
records <- get_previews(records)
view_previews(records[5:8, ])
```

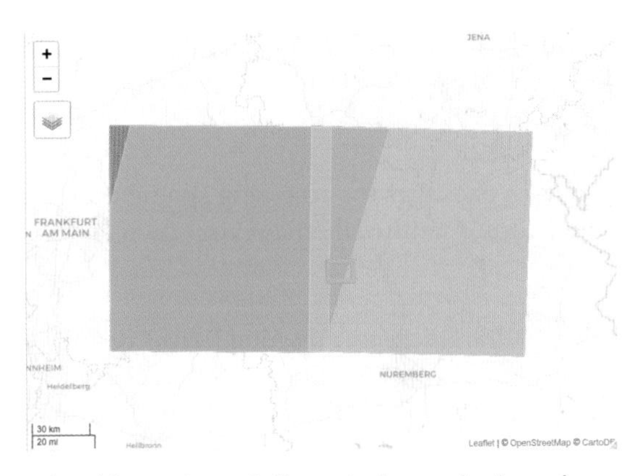

Figure 12.15 Geographical footprints of all queried records viewed on an interactive map.

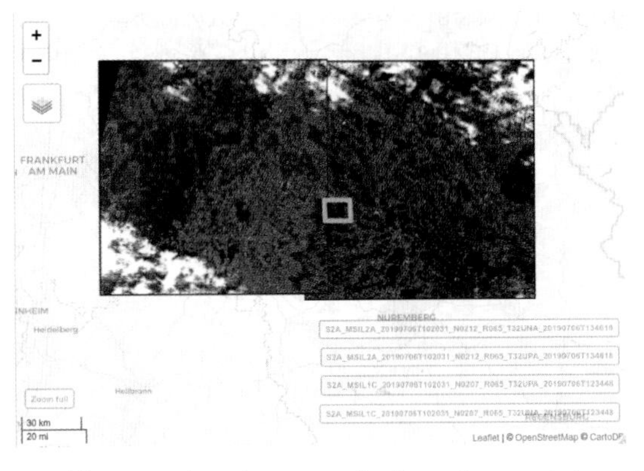

Figure 12.16 Previews of four overlapping records viewed on an interactive map. Previewing allows you to visually assess the usability of a dataset, for example, by checking cloud cover.

Instead of visually checking each preview image, you can filter records, for example, by cloud cover. In the following example, records with an overall cloud cover <20% are kept whereas all records with >20% cloud cover are removed:

```
records <- records[records$cloudcov < 20, ]
```

getSpatialData can also automatically analyse each preview of a record to select data sets with low or no cloud coverage inside your AOI. If you are interested in these features, have a look at the help of the `select` function by calling `?select`.

To finally download the selected records, use `get_data()`. Then, import the data into R as a list of *raster* objects using `read_data()`:

```
records <- get_data(records)
imagery <- read_data(records)
```

TASK

Practise with the functions of the *getSpatialData* package. Query the records of different sensors. Try to filter and select those records you want to use for an analysis (e.g. no or only a few clouds) before downloading them. *getSpatialData* can help to remove records that are not useful for your analysis before download.

There are many different packages for accessing spatial data from different sources. If you are interested in exploring some of them on your own, we recommend you have a look at the following packages:

- *rnaturalearth* and *rnaturalearthdata* to download free, openly available vector and raster data from *Natural Earth* (http://naturalearthdata.com).
- *osmdata* for collecting vector data from OpenStreetMap (https://www.openstreetmap.org/#map=6/54.910/-3.432).

12.5 Organization of spatial analysis scripts

If you have not yet started to use single files for individual analyses, we recommend doing so from now on. In the long run, it will be beneficial for you (and your collaborators) if you can easily re-execute any analysis. We also recommend writing analyses as generic as possible and storing them in your personal script repository. The names of single scripts should also be self-explanatory. This is also true for the names of variables. Do not use generic names since you will find it much harder to read and understand your code after a while. Use comments in R to provide additional information for each spatial script, especially concerning data formats:

```
# Short description of what the script does
# ############
# Purpose: This is a script to import GPS data and convert it
to SHP format. Data sets will be converted ...
# Author: your name
```

```
# Date: month, year
# R version and packages: x.xx
# ##############
# ##############
# Import format: csv (tab-delimited)
# Output format: GeoPackage
# ##############

# Load required packages
library(sp)

# Import the CSV file with x and y coordinates
gps_in <- read.table("/path/to/mydata.csv")

# Coordinate columns ought to be named ....

coordinates(gps_in) <- ....

# #### Data integrity check

# Check first entries
head(gps_in)

# Overview statistics
summary(gps_in)

# Plot data
plot(gps_in)

# Data analysis
...

# Data conversion for further analysis using sf
...

# Export as GeoPackage format
st_write(...)
```

12.6 Summary

This chapter introduced you to the basic spatial data handling functionalities in R. Many more options exist, and you will encounter a variety of hurdles. However, constantly working within this environment and having a specific goal in mind will enable you to understand and implement many spatial operations for your project using the R command line.

13. Spatial analysis in R

The options to conduct spatial data analysis in R are many; due to the nature of open-source software, the source codes of all methods used can be accessed and modified. In the following sections, we introduce several fundamental spatial analysis functions to empower you to conduct your own analysis on the command line.

13.1 Vegetation indices

Calculating indices within R is probably easier than using QGIS, simply because you can store and reuse any function. Even though various functions calculate widely known indices from raster data automatically, it is easy to implement the formula of a specific index in R and apply it to all pixels of a raster. The *raster* package makes mathematical operations on rasters simple. In this example, we multiply the values of the first band in our raster stack by 10:

```
s2_b1x10 <- s2[[1]] * 10
```

This is equivalent to using the Raster Calculator in QGIS. Further operators can also be implemented, for example, to calculate the normalized difference vegetation index (NDVI) using the third and fourth band in our raster stack:

```
ndvi <- (s2[[4]] - s2[[3]])/(s2[[4]] + s2[[3]])
```

The result is a single-layer raster, like the one derived from using the Raster Calculator in QGIS. Calling `summary()` on the new object is good practice just to check whether the value range of NDVI (–1 to +1) is correct. Plotting the object using the *raster* plot function will result in a similar map to the one we have already seen in QGIS (Figure 13.1):

```
plot(ndvi)
```

As the plot key shows, dark green areas indicate high NDVI values whereas yellowish and brownish colours are associated with lower values representing bare soil or settlements. These data can be used for further analysis, but it is also good practice to compare the vegetation index to the output of other indices because all provide slightly different results. Other indices, such as the modified soil-adjusted vegetation index (MSAVI), can be calculated similarly. However, this requires entering more complex formulas, which might be time-consuming. The *RStoolbox* package provides a function that allows us to compute many implemented indices, such as the NDVI or MSAVI, using the near-infrared (NIR) and RED reflectance layers automatically:

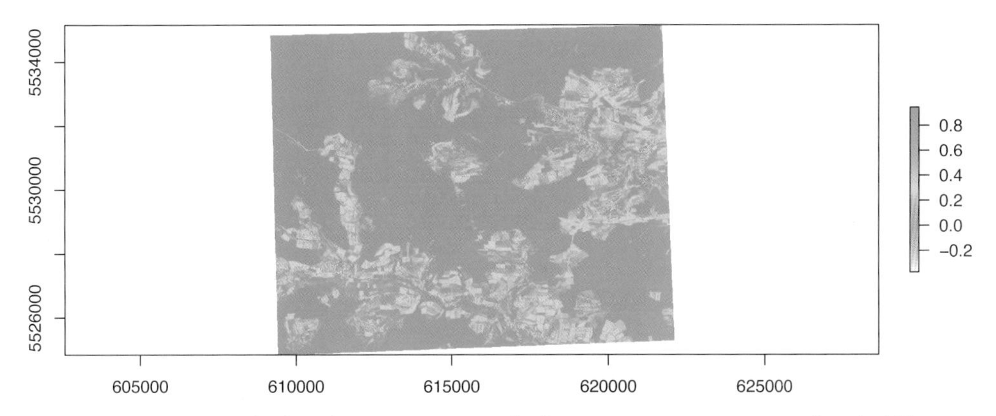

Figure 13.1 NDVI calculated using raster calculation in R. NDVI, normalized difference vegetation index.

```
library(RStoolbox)
ndvi <- spectralIndices(s2, red = 3, nir = 4, indices =
"NDVI")
```

Because the function does not know which spectral bands are in which position in your raster object, you must define the layer number for each spectral band. This function results in the same NDVI output as shown before. This, as well as all other raster objects, regardless of whether they are single or stacked objects, can be saved to your hard drive using:

```
writeRaster(ndvi, "ndvi_S2.grd")
```

A meaningful name, which also indicates the origin of the analysis, is highly recommended. Thus, with more complex scripts and multiple analyses, it is better to use the full original Sentinel filename plus the additional information from the calculation of the NDVI so that NDVI calculations using different data sets are not confused.

Omitting the `indices="NDVI` argument definition results in a large raster stack with all indices using the defined bands being calculated:

```
indices <- spectralIndices(s2, red = 3, nir = 4)
```

This analysis results in a 13-band raster brick, all being vegetation indices using the RED and NIR bands. We can also limit the number of indices being calculated by specifically listing the indices that should be calculated (Figure 13.2):

```
indices <- spectralIndices(s2, red = 3, nir = 4, indices =
c("CTVI", "DVI",
    "GEMI", "MSAVI", "MSAVI2", "NDVI"))
```

Plotting the resulting raster object shows the diversity of all six indices:

```
plot(indices)
```

Calculated indices such as the NDVI, MSAVI or global environment monitoring index show a discrepancy concerning landscape patterns and value ranges. Generally, all

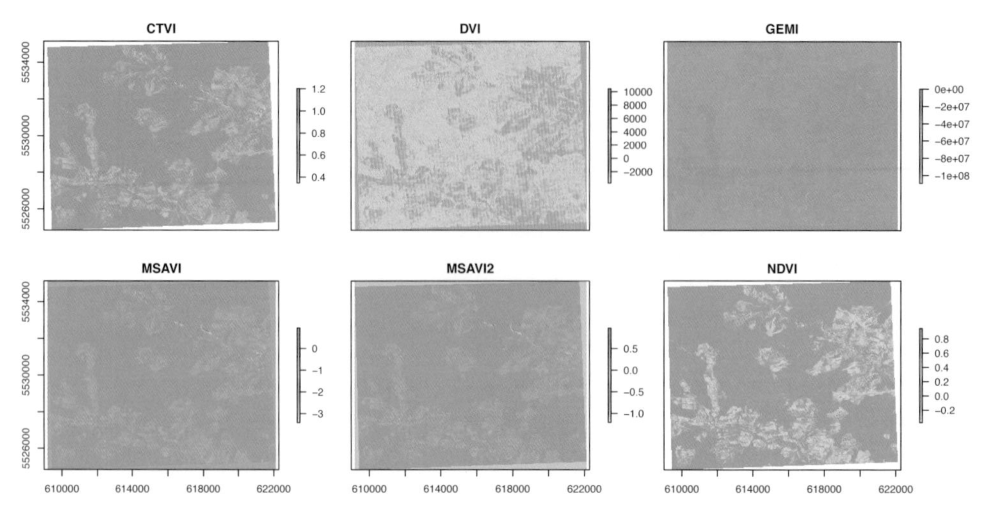

Figure 13.2 Several spectral indices using the RED and NIR bands are calculated and plotted using the `spectralIndices` function. CTVI, corrected transformed vegetation index; DVI, difference vegetation index; GEMI, global environmental monitoring index; MSAVI, modified soil-adjusted vegetation index; NDVI, normalized difference vegetation index.

indices show the same pattern but with subtle differences that might be relevant for your study. Therefore, it is good practice not to rely on one index only but to also check or evaluate others. Moreover, assigning further bands would result in even more indices that use the newly defined bands being calculated, for example, blue or green bands.

> **TASK**
>
> Add other spectral band definitions in `spectralIndices()` and recalculate the indices. Compare indices that use different spectral information.

13.2 Digital elevation model (DEM) derivatives

Deriving elevational information through the Shuttle Radar Topography Mission DEM is also possible in R. Because we then want to combine the DEM derivatives with the Sentinel-2 data, we need to resample the DEM data to the Sentinel data. First, we must import the file. Because it consists of only one layer, we can use `raster()` instead of `brick()`:

```
dem <- raster("N49E010_UTM_WGS84_32N_crop.tiff")
```

The DEM resolution is 78 m while the Sentinel-2 is 10 m (information retrieved by executing their name); thus, we must resample the Sentinel data to 78 m or vice versa. Because we want to keep the 10-m resolution for further analysis, we change the resolution of the DEM data set even though doing so introduces a pseudoreplication of DEM values within the 10-m resolution data (Figure 13.3):

```
dem_resampled_10m <- resample(dem, s2)
```

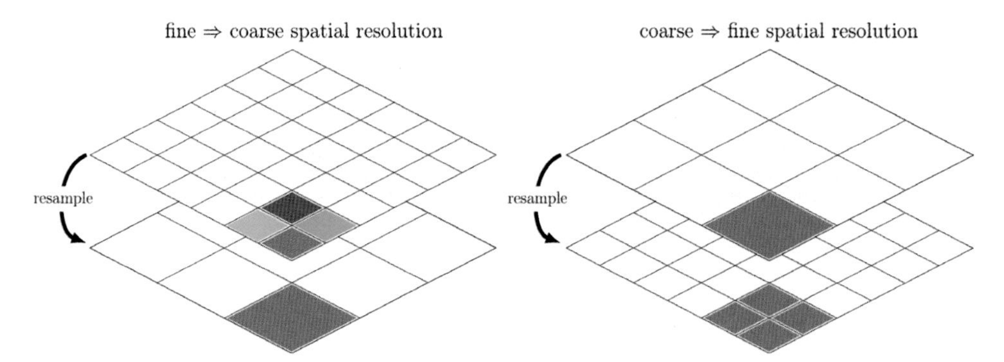

Figure 13.3 Implications of resampling raster values from fine to coarse spatial resolution and vice versa. From coarse to fine, the same value is replicated for all matching high spatial resolution pixels.

Now, we could stack both data sets together and use them for further analysis. However, we first calculate certain terrain features, such as slope and aspect:

```
dem_slope <- terrain(dem_resampled_10m, opt = "slope")
dem_aspect <- terrain(dem_resampled_10m, opt = "aspect")
dem_roughness <- terrain(dem_resampled_10m, opt = "roughness")
```

The new rasters can then be plotted and display the slope, aspect or roughness of our study area. If you wanted to save them for later use, all files must be written to your hard drive using `writeRaster()`. This can also be done after all three raster objects are stacked together so that the whole stack is exported.

> **TASK**
>
> Try the `mapview()` function in the *mapview* R package to display your data in an interactive mapping window.

13.3 Classification

In previous chapters, we discussed classifications in general and in QGIS specifically. Conducting them in R is comparably straightforward. Calculating both *unsupervised* and *supervised* classifications can be done by using the *RStoolbox* package.

13.3.1 Unsupervised classification

For *unsupervised* classifications, *RStoolbox* provides the `unsuperClass()` function, which has the required functionality:

```
unsup_studyArea <- unsuperClass(s2, nClasses = 5)
```

The result is a classified raster of our study area with five classes based on *k*-means clustering. The output can be displayed using a slightly different syntax because the

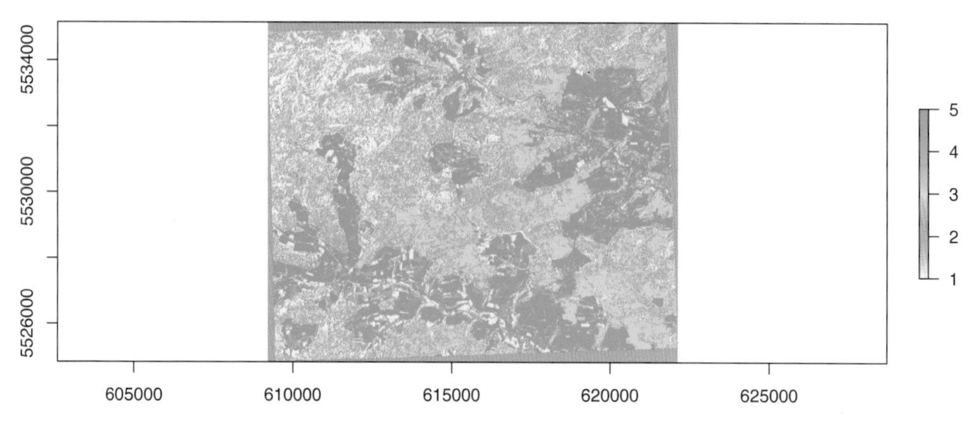

Figure 13.4 Unsupervised classification of our study area.

resulting object is not only a raster but also holds further information. Therefore, we must specifically address the raster map:

```
plot(unsup_studyArea$map)
```

This map (Figure 13.4) shows discrete values that represent spectrally different classes defined by the *k*-means algorithm and the number of classes defined a priori. The legend is misleading because it suggests gradual values being displayed in the map. This issue can be resolved using the *ggplot2* mapping syntax for spatial data, which is introduced in Chapter 15.

Any classification can be improved by including more layers containing relevant environmental information, such as NDVI or terrain slope. The data sets need to be stacked before running the classification again:

```
raster_stack <- stack(s2, ndvi, dem_slope, dem_aspect,
dem_roughness)
```

Now, we can execute the unsupervised classification again, this time using the stacked raster object:

```
unsup_studyArea_2 <- unsuperClass(raster_stack, nClasses = 5)
```

We can now display the resulting map and compare it to the results of the previous classification. Changing the number of classes and executing the script again will change the result. Unsupervised classifications are meant for explorative analysis of landscape feature, to discriminate between features, or for preliminary landscape classifications.

13.3.2 Supervised classification

In contrast, *supervised* classification approaches require training data stored as vector files. Training polygons are created before the analysis through on-screen digitization, usually using QGIS, and must be imported beforehand:

```
library(sf)
train_samples <- st_read("train_samples.shp")
```

This layer is a vector polygon with several attribute columns. The most important one for us is the 'class' column. Moreover, it is good practice to check whether the projection, especially of our raster data, matches that of our chosen projection. Calling the vector object by typing its name shows that the projection is 'longlat' while the Sentinel-2 data are 'UTM'. The vector data must be in the same projection as the Sentinel data, otherwise they cannot be combined within the classification operation; thus, it must be reprojected:

```
train_samples_utm <- st_transform(train_samples,
st_crs(raster_stack))
```

Generally, polygons should cover the whole study area to represent land cover variability within the landscape. For our analysis, we use the vector object together with the raster stack in the supervised classification function provided by *RStoolbox*. Note that we first convert the training polygons from *sf* to *sp* since *RStoolbox* expects an *sp* object as input. Calling `superClass()` will likely result in a message that further packages, such as *randomForest*, are required. Install these packages and re-execute the classification function:

```
train_samples_utm.sp <- as(train_samples_utm, "Spatial")
sc_studyArea <- superClass(raster_stack, trainData = train_
samples_utm.sp, responseCol = "class")
```

The response column defines which information is used for the actual land cover classes. The column for the classes can be numeric or character-based; both work with the `superClass()` function. Plotting the supervised classification (Figure 13.5) shows three classes in our landscape; this is due to using a training input consisting of three categories:

```
plot(sc_studyArea$map)
```

This classified raster already shows the pattern of our landscape. Most of the landscape is covered by forest, which we defined as class 1. Certain misclassifications can be identified and could be improved by adding either more training data or further environmental layers that contribute additional relevant information, which can be used to discriminate classes. You can visualize the classification result using the interactive mapping option,

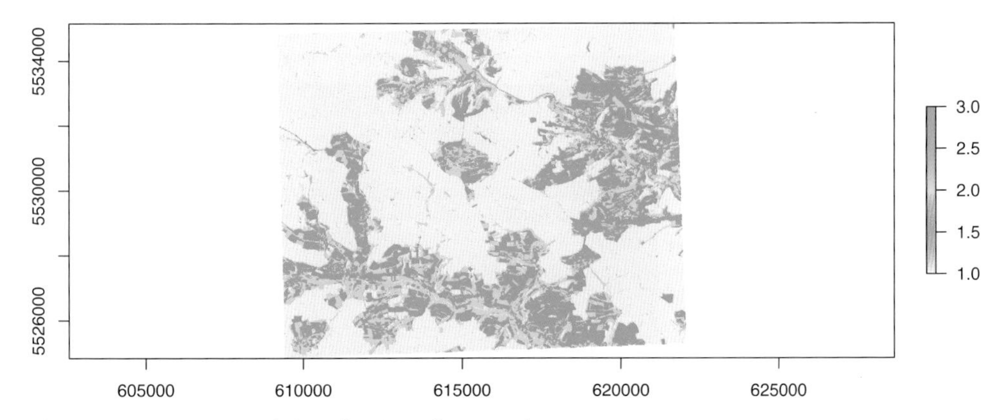

Figure 13.5 Supervised classification of our study area.

which allows you to zoom and browse through the map, thus making the visual evaluation easier:

```
library(mapview)
mapview(sc_studyArea$map, col.regions = c("darkgreen",
"lightgreen", "yellow"))
```

For any analysis, we need to provide measures of accuracy. As we explained in the QGIS classification part of the book (Chapter 9), two options exist: either the training data is automatically split into training and validation samples, or a totally independent validation polygon layer is provided. Both approaches provide information about accuracy, such as overall accuracy or user and producer accuracy, as described in Chapter 9 (classifications using QGIS). In R, the words *precision* and *sensitivity* are used (Figure 13.6). However, to retrieve accuracy measures, we must run the classification again and split the training data into training and validation data sets by setting the argument `trainPartition=0.7`, which allocates 70% to training and 30% to validation:

```
sc_studyArea <- superClass(raster_stack, trainData = train_
samples_utm.sp, responseCol = "class",
    trainPartition = 0.7)
```

Overall accuracy and further metrics can be extracted from the resulting file by calling the object or executing `getValidation()`. Validating an existing map using your own validation data can be achieved by using the `validateMap()` function provided by *RStoolbox*. Providing the information about accuracy together with your final land cover map is important to judge the quality of the classification. In our example, the overall accuracy is quite high; for more complex landscapes, this is rarely achievable. As you can see from the confusion matrix (the table at the beginning with the assigned predicted and reference classes) forest is always correctly mapped but some confusion exists between meadow and soil. This is also reflected by individual class accuracy. The resulting classification should now be saved to your hard drive by:

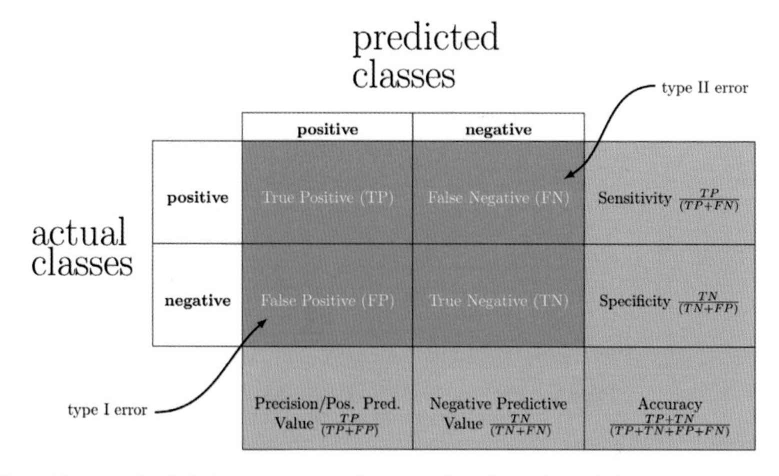

Figure 13.6 Matrix of validation measures for actual and predicted classes.

```
writeRaster(sc_studyArea$map, "supervised_classification_
Study_Area.grd")
```

Please note that, using `$map`, you have exported only the actual map and not the whole supervised classification object.

13.4 Raster-vector interaction

In the following example, the landscape analysis based on the raster data is linked to points within the study area. Polygons and lines could be used if necessary. Extracting the NDVI value for our settlement locations can be done after importing them into our R session:

```
places <- st_read("place_UTM_WGS84_32N_clip.shp")
```

These points are then used to query the NDVI values at the very same coordinates and are assigned to a new object. Since `extract()` requires an `sp` object, we must convert `places` from an `sf` to an `sp` single-point data frame. Thus, we use `st_cast()` to create single points from multiple points:

```
extract_ndvi_poi <- extract(ndvi, as_Spatial(st_cast(places,
to = "POINT", warn = FALSE)))
```

The content is a list of numbers and can be plotted as a non-spatial plot (Figure 13.7) with the NDVI values on the y-axis and the location index on the x-axis:

```
plot(extract_ndvi_poi)
```

The mean and variance within a polygon can be retrieved, for example, if we wanted to know whether a defined environmental value was equally distributed within the study polygons. Staying with the current example of points, we want to create a 200-m buffer around the points (Figure 13.8):

```
buffer_polygons_200m <- st_buffer(places, dist = 200)
```

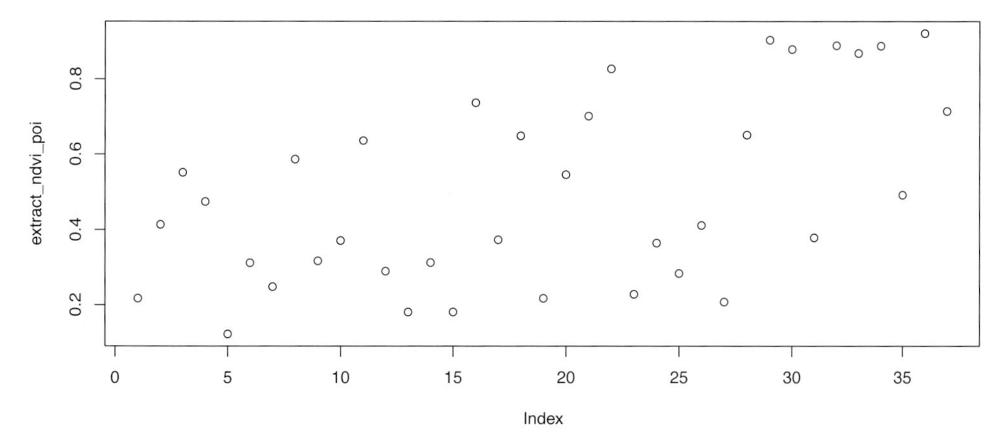

Figure 13.7 Scatterplot of the NDVI values (y-axis) for each location index (x-axis).

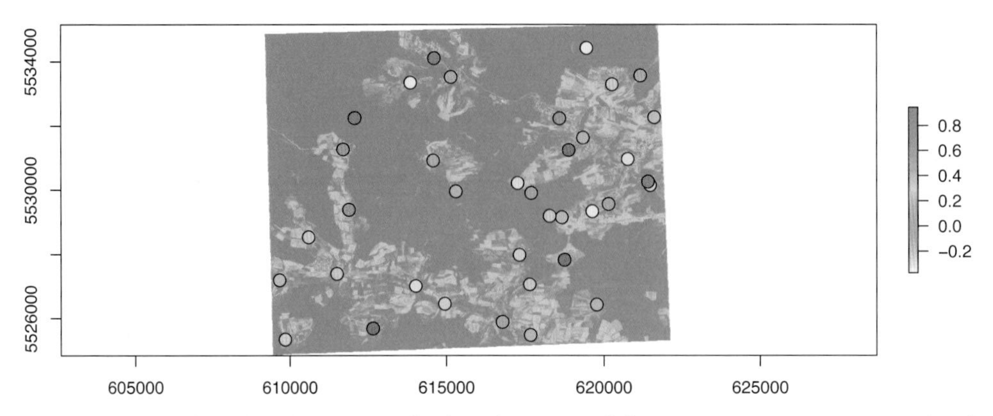

Figure 13.8 Buffer of each location displayed on top of the NDVI raster. NDVI, normalized difference vegetation index.

The resulting object can be displayed spatially on top of the original point locations or any raster, such as the NDVI, using the `add=T` option:

```
plot(ndvi)
plot(buffer_polygons_200m, add = T)
```

However, in our example, we want to query the values of a range of environmental attributes. This requires that we first stack all environmental raster objects that we want to retrieve:

```
raster_stack_vect <- stack(ndvi, dem_slope, dem_aspect, dem_
roughness, sc_studyArea$map)
```

Extracting the cell values within polygons, our newly created buffer areas, requires a few intermediate steps. First, we must extract the raster cell numbers within the buffer:

```
buffer_cells <- cellFromPolygon(raster_stack_vect,
as(buffer_polygons_200m,
    "Spatial"))
```

This returns a list that is as long as the number of buffer polygons. Each element in the list is a numeric vector that contains the number of the cells covered by the respective polygon. From these cell numbers, which indicate the cell position on the raster, we can extract the corresponding pixel values per polygon. We do this by looping through the list of vectors using the `lapply()` function. Each vector in `buffer_cells` (one after another) is used to query the raster stack. The result of `lapply` is again a list as long as `buffer_cells`, which does not contain cell numbers, rather pixel values per polygon:

```
buffer_values <- lapply(buffer_cells, function(x)
raster_stack_vect[x])
```

Now, we have all the values within our buffer areas; however, since we extracted the values as simple numeric vectors, we have lost the spatial information. If we want to work with a spatial vector file holding the environmental raster attributes, then we must run a few more functions. First, we need to query the coordinates for the cells we extracted the values from:

```
buffer_points <- lapply(buffer_cells, function(x)
xyFromCell(raster_stack_vect,
    x))
```

Then, we can join the point coordinates and cell values to create a spatial object. We use a loop function like `lapply()`, called `mapply()`. The difference is that `mapply()` can loop through several objects of equal lengths at the same time. In this example, the first element of each polygon's name, corresponding cell coordinates and pixel values are selected and bound together as columns to a data frame using `cbind.data.frame()`. Then, the task is iterated on the second element of these three objects, then the third, fourth, and so on. When finished, we have a list of data frames. Each data frame represents one polygon and contains its name, the corresponding cell coordinates on our raster grid and the associated pixel values.

```
buffer_df <- mapply(name = buffer_polygons_200m$name, coords =
buffer_points,
    val = buffer_values, cbind.data.frame, SIMPLIFY = F)
```

Since we do not want to have multiple separate data frames but only one, we can use `rbind.data.frame()` to bind all data frames together as subsequent rows. Using the `do.call()` function, we take all elements of `buffer_df` (all are data frames) and supply them to `rbind.data.frame()`:

```
buffer_df <- do.call(rbind.data.frame, buffer_df)
```

At the outset, it may be difficult to understand these lines of codes. We recommend you read the manuals of `lapply()`, `mapply()` and `do.call()` to get a more detailed understanding of how they work. Even though the code looks complicated at first glance, we need only five lines of code to produce a highly informative result, which can be repeated using any set of polygons or raster objects, that is, a data frame with the polygon name, its coordinates and all environmental values. We can now make some minor adjustments, such as shortening the attribute names or exchanging class names and numbers:

```
colnames(buffer_df)[4:ncol(buffer_df)] <- c("v_NDVI", "v_slp",
"v_asp", "v_roug",
    "v_class")  # Shorten names
buffer_df$v_class <- plyr::mapvalues(buffer_df$v_class, c(1,
2, 3), c("forest",
    "meadow", "soil_rock"))  # Map names to class numbers
```

Finally, the spatial object can be created and plotted as such (Figure 13.9):

```
buffer_sf <- st_as_sf(buffer_df, coords = c("coords.x",
"coords.y"))
plot(buffer_sf)
```

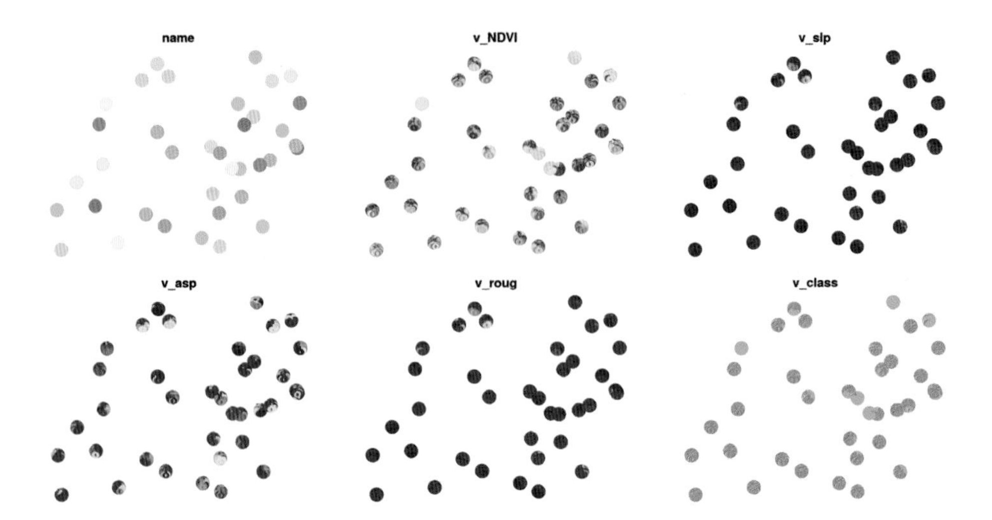

Figure 13.9 Environmental values per pixel for each buffer polygon.

The plot shows the single values for each environmental parameter inside each buffer. There are other ways to achieve this result, for example, by masking raster values based on the vector buffer areas using the `mask()` function, but then we would have a raster output instead. Finally, we save the spatial vector object we have created to our hard drive into the *data* folder for later use; such a folder must be located on your hard drive:

```
st_write(buffer_sf, "data/places_buffer_points.shp")
```

If you have already saved the data to your hard drive using the same name and location, you must overwrite it using `delete_dsn=TRUE` within the `st_write()` function.

13.5 Calculating and saving aggregated values

The example used in the preceding sections worked with all values within buffer areas. However, in certain situations, aggregated values are preferred. This process is more complex, especially if we also want to extract the median, standard deviation and the number of classes within a buffer. Thus, we want to add several statistical calculations within the same process. However, we can create and apply a customized function that extracts the statistical values and outputs a list of values for each polygon with aggregated information:

```
# Create function
measures_fun <- function(x) {

    # Calculate measures for continuous variables and assign
name
    c_mean <- apply(x[, 1:4], MARGIN = 2, mean)
    names(c_mean) <- paste0("mean_", names(c_mean))
```

```
    c_median <- apply(x[, 1:4], MARGIN = 2, median)
    names(c_median) <- paste0("median_", names(c_median))

    c_sd <- apply(x[, 1:4], MARGIN = 2, sd)
    names(c_sd) <- paste0("sd_", names(c_sd))

    # Calculate measures for discrete class variables
    n_classes <- c(n.forest = length(which(x[, 5] == 1)),
n.meadow = length(which(x[,
        5] == 2)), n.soil = length(which(x[, 5] == 3)), n.all
= length(x[, 5]))

    # Return a numeric vector of mean, median, sd and n. of
classes
    return(c(c_mean, c_median, c_sd, n_classes))
}

# Apply function
buffer_measures <- sapply(buffer_values, measures_fun,
simplify = F)
```

The last steps change the formatting of this information into a data frame that results in locations being listed as rows and statistical values being listed as columns. We rename the column names, otherwise the export function to SHP format, for example, will shorten the names arbitrarily:

```
buffer_measures <- do.call(rbind, buffer_measures)
colnames(buffer_measures) <- c("m_NDVI", "m_slp", "m_asp", "m_
rough", "md_NDVI",
    "md_slp", "md_asp", "md_roug", "sd_NDVI", "sd_slp", "sd_
asp", "sd_roug",
    "n_forst", "n_meadow", "n_soil", "n_all")
```

The values are then combined with the buffer polygons into a spatial object that contains the summary statistics for each buffer:

```
buffer_sf_aggr <- cbind(buffer_polygons_200m, buffer_measures)
```

We can plot this; a map for each single column entry is displayed:

```
plot(buffer_sf_aggr)
```

This spatial object can also be saved to our hard drive so we can use it in Chapter 14:

```
st_write(buffer_sf_aggr, "data/places_buffer_aggr_polygons.
shp")
```

Both spatial data sets will be used in Chapter 14 to display their data in a non-spatial way. Exporting such objects is a common task because it allows us to use these data in different analyses as well as sharing them with colleagues. Thus, all exported or saved data must be named correctly and meta-information (and the code used to create it) must be provided to allow fellow researchers to understand the data structure and their content. Otherwise, data might not be used or worse, it may be interpreted wrongly.

13.6 Summary and further reading

In this chapter, you learnt the basics of conducting spatial analyses in R, including calculating indices, classifying data and scripting vector-raster queries. We have covered commonly used operations like `crop()` and `mask()`. Other functions might be relevant for your analysis, such as `projectRaster()` to reproject a raster or `resample()` to change the spatial resolution. The distance analysis we performed with QGIS can also be achieved using the `distance()` function in R. Converting raster to vector or vice versa can be achieved using the `rasterize()` or `rasterToPolygon()` functions. Many more functions are available in the packages we have covered as well as other packages, such as *rgeos*, *maps* or *lwgeom*. For some analyses, landscape fragmentation might also be relevant; the *landscapemetrics* package can be used for this, as well as the *NLMR* or *belg* package. Currently, *sp* to *sf* conversion is still needed for certain tasks; ongoing development of *sf* will continue to reduce such a need. We highly recommend you keep up to date with the development of packages in R that deal with spatial data to learn about new features that might make working with spatial data in R easier.

14. Creating graphs in R

When working in the spatial realm, we primarily focus on map making; however, some information cannot be shown using maps but requires graphs instead. For example, showing box plots of normalized difference vegetation index (NDVI) values across all field sites or ecological regions is not possible using a map. Such non-spatial graphs can be very informative and important for further analysis or interpretation.

At first, creating graphs in R does not seem to be a pragmatic way to create a figure; however, it is worthwhile in the long run and graphs can be sophisticated and look elegant. Additionally, you can easily apply the R code to any other data or to new versions of your data. Moreover, carrying out all steps in one environment, that is, data management, analysis and plotting, allows you to quickly adapt your code and data and reduces potential errors due to, for example, export and import routines, which might compromise data structure.

You can create graphs and maps with the default packages provided by R; however, for more sophisticated graphs and maps, you need some additional packages. Regarding spatial mapping, R's default capabilities are quite limited. Therefore, in this chapter, we introduce the *ggplot2* package and start with some non-spatial graphs before delving into maps with *ggplot2*.

We use the vector data created in Chapter 13 and start with the aggregated environmental information before we use non-aggregated values.

14.1 Aggregated environmental information

First, we need to import our data set if it is not already in our R environment:

```
library(rgdal)
places <- readOGR("places_buffer_aggr_polygons.shp")
```

The imported data set is a spatial vector object; thus, we first must extract the data from our spatial object before we can use it to create non-spatial graphs. As we explained previously, data inside a spatial vector are structured as a normal data frame; therefore, they can be easily extracted and used for common non-spatial operations:

```
df <- places@data
```

If you want to check whether you are dealing with the same data as in Chapter 13, simply compare the data frame we used previously with the information extracted from within the spatial vector object. Using the `names()` or `summary()` functions provides us with initial information about column names and summary statistics, respectively. In the following sections, we want to display the summary statistics for the 14th to 17th column:

```
summary(df[, 14:17])

#>      m_NDVI            m_slp              m_asp
m_rough
#> Min.   :0.4555   Min.   :0.02562   Min.   :1.143   Min.
:0.6274
#> 1st Qu.:0.4970   1st Qu.:0.05872   1st Qu.:2.542   1st
Qu.:1.4751
#> Median :0.5556   Median :0.07324   Median :2.731   Median
:1.8989
#> Mean   :0.5977   Mean   :0.07822   Mean   :2.792   Mean
:1.9591
#> 3rd Qu.:0.6541   3rd Qu.:0.09789   3rd Qu.:3.010   3rd
Qu.:2.4418
#> Max.   :0.9096   Max.   :0.15493   Max.   :4.757   Max.
:3.9191
```

This information tells us the minimum, maximum or mean values of certain columns (mean of the NDVI, mean slope, and so on). More importantly, it allows us to check whether the columns were properly created and if the value ranges make sense. As mentioned in Chapter 13, we recommend running a first check on the data each time to verify that the data are properly imported. Apart from base functions such as `summary()`, `head()` or `tail()`, dedicated 'explorative data analysis' packages, such as *dataMaid*, can be used for further preliminary data inspection. Plotting your data is always a good first explorative approach.

We have already used functions such as `plot()`, but this function has its limits. While the *ggplot2* syntax is quite different and might look complex initially, it has its advantages. It allows us to combine several plot functions with the + sign. Each function has its own meaning and produces a certain plot type. The key function we need to use is `ggplot()`; however, this function does not specify any further details of our plot, such as the type of plot (e.g. point or box plot) we want to use. We look at this syntax by simply plotting a scatter plot of our data frame *df* using `names` against the mean NDVI `m_NDVI` (Figure 14.1):

```
library(ggplot2)
ggplot(df, aes(name, m_NDVI)) + geom_point()
```

The `aes()` function within `ggplot()` is used to define what is plotted on the x- and y-axes. The second part, `+ geom_point()`, defines the type of plot. The function can of be adapted further to consider additional statistical values, for example, the standard deviation of the NDVI, as point sizes (Figure 14.2):

```
ggplot(df, aes(name, m_NDVI, size = sd_NDVI)) + geom_point(col
= "darkgreen")
```

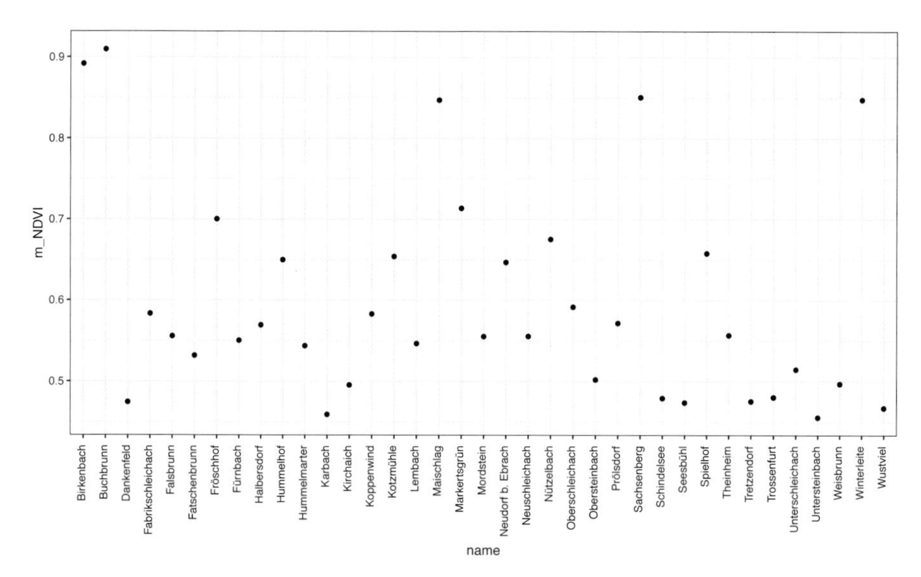

Figure 14.1 Simple scatter plot of mean NDVI values using default settings, resulting in overlapping axis labels that need to be adjusted to make the plot easier to understand. NDVI, normalized difference vegetation index.

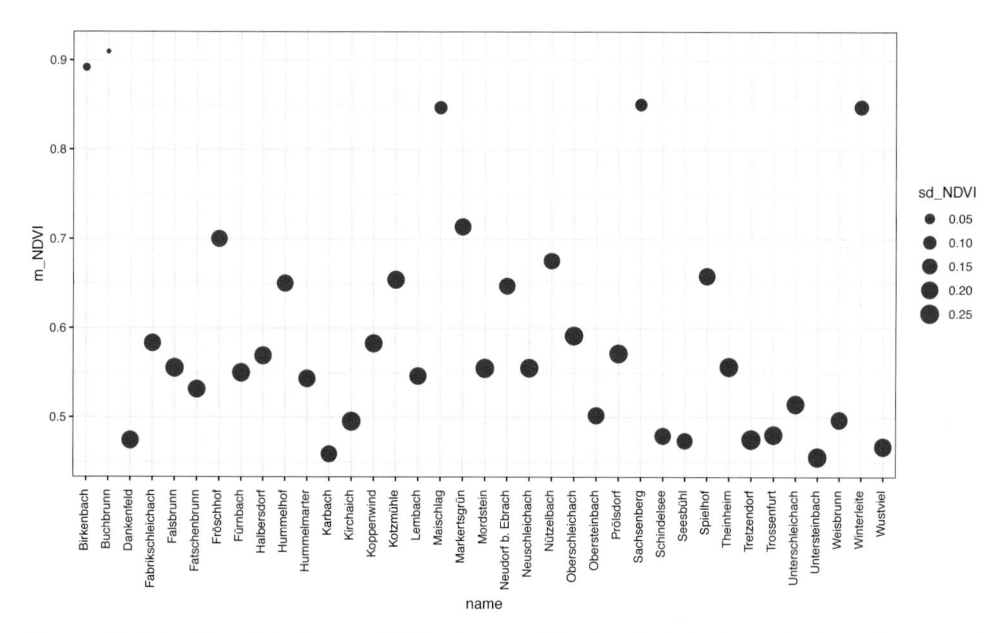

Figure 14.2 Scatter plot of mean NDVI values and its standard deviation indicated by size of the dots. NDVI, normalized difference vegetation index.

> ### ✍ TASK
>
> Change the values of the column names used for x, y, size or colour of the plots, respectively and check how your plot changes.

The colour of points can be changed using the `col=` definition within the `geom_point()` function. Other parameters can also be adapted, such as translucency using `alpha=`. We can also colour each point individually based on the attributes of a column value (Figure 14.3):

```
ggplot(df, aes(name, m_NDVI, size = sd_NDVI, col = n_forst)) +
geom_point()
```

Further options, such as the shape of the points, can also reflect a column attribute. Additionally, labels and legend names can be adapted. This is covered at the end of this chapter, when we finalize the plots and save them.

> ### ✍ TASK
>
> Have a look at the *rayshader* package (https://github.com/tylermorganwall/rayshader) and display your *ggplot2* graph three-dimensionally.

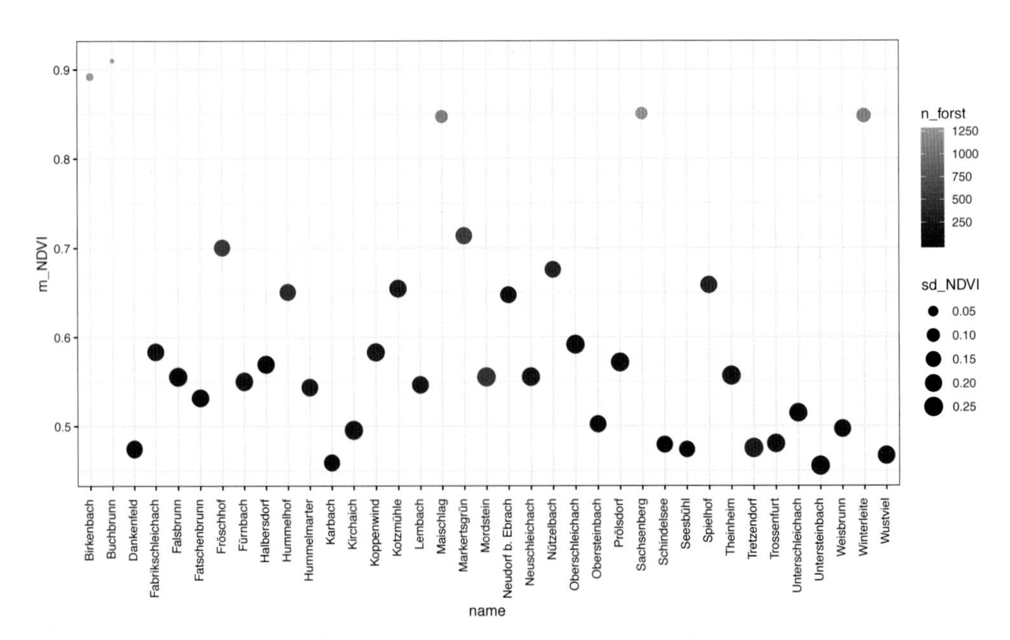

Figure 14.3 Mean NDVI values shown in a scatter plot with changing dot sizes based on the standard deviation; colour is defined by the number of forest pixels. NDVI, normalized difference vegetation index.

14.2 Non-aggregated environmental information

The second data set created in Chapter 13 contains non-aggregated environmental information. These data provide the same information as the previous data frame; they contain all single data points and no aggregated statistics. The process is the same, but we have more options to display such data. However, we must first import the data set:

```
library(rgdal)
places <- readOGR("places_buffer_points.shp")
df <- places@data
```

Displaying the imported data set works as explained earlier, but we can apply different and more complex plot functions, such as a box plot (Figure 14.4), because we have more than one value per location:

```
ggplot(df, aes(name, v_NDVI)) + geom_boxplot()
```

The box plot shows that certain villages have a wide range of NDVI values whereas others have a narrow range with lower or higher values only. This gives us an idea as to which villages are generally 'green', have hardly any green spaces or show high variation. The dots indicating any outliers, while the horizontal lines within the boxes show the median (Figure 14.5).

In the previous plots, it was impossible to read the names of locations because they overlap one another. However, there are solutions to this problem: we can rotate the names by 45 or 90 degrees using `+ theme(axis.text.x = element_text(angle = 45, hjust=1));` or we can flip the whole plot (Figure 14.6):

```
ggplot(df, aes(name, v_NDVI)) + geom_boxplot() + coord_flip()
```

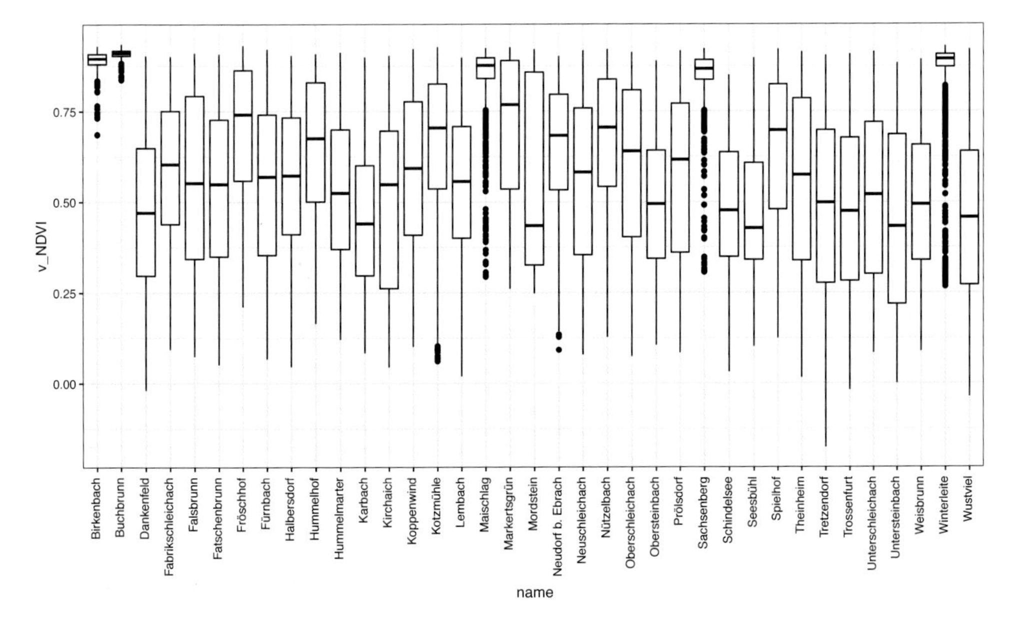

Figure 14.4 Simple box plot of all NDVI values.

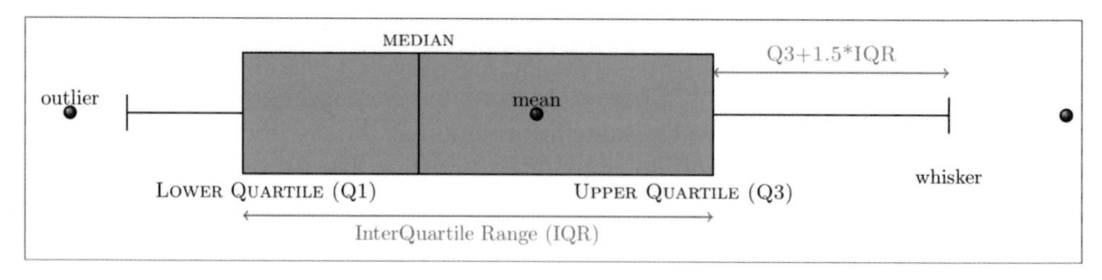

Figure 14.5 Explanation of box plot structure and interpretation of displayed data.

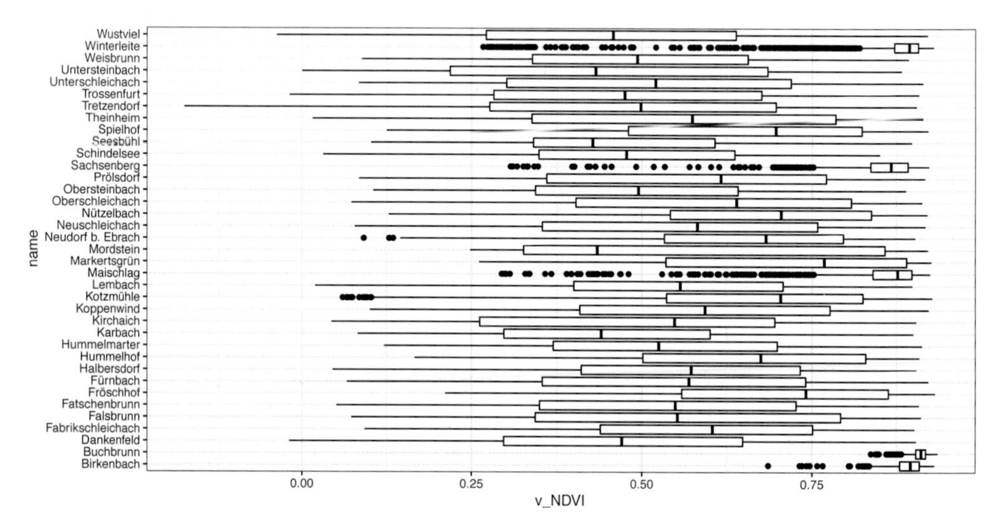

Figure 14.6 Simple box plot of all NDVI values but with flipped x- and y-axes.

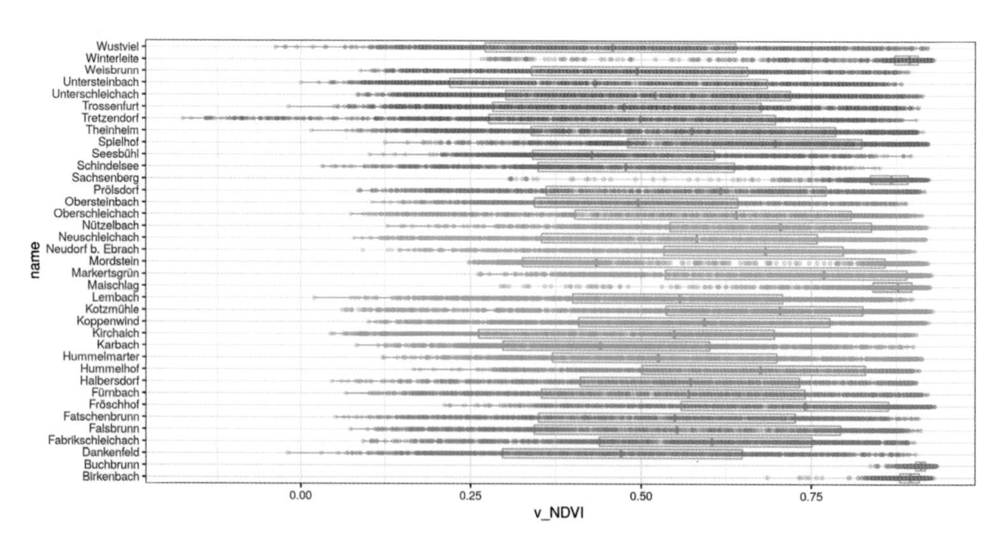

Figure 14.7 Box plot of NDVI values with adapted size and colour settings. NDVI, normalized difference vegetation index.

Now, we can read the location names easily, but the plot still needs tidying up. Thus, we define that each location should have a separate colour (`col=name`) and add points for each value on top of the box plot (`+geom_point`). This also allows us to better interpret the actual pattern (Figure 14.7):

```
ggplot(df, aes(name, v_NDVI, col = name)) + geom_boxplot(color
= "grey", outlier.shape = NA) +
    coord_flip() + geom_point(size = 1, alpha = 0.2) +
theme(legend.position = "none")
```

Functions are partly self-explanatory: `alpha` defines translucency; `size` defines point size; and `theme()` tells R that no legend should be added. This makes sense since colours are defined by the names of locations, which are already listed on the left axis. To interpret the plot about the difference between single location mean values and the general mean, we can add a line indicating the mean across all locations. First, we must calculate the mean:

```
ndvi_mean <- mean(df$v_NDVI)
```

Then, we can use this value within `geom_hline()` as the `yintercept` value and define a specific size, translucency and line type (Figure 14.8):

```
ggplot(df, aes(name, v_NDVI, col = name)) + geom_boxplot(color
= "grey", outlier.shape = NA) +
    coord_flip() + geom_point(size = 1, alpha = 0.2) +
theme(legend.position = "none") +
    geom_hline(aes(yintercept = ndvi_mean), color = "black",
size = 0.8, alpha = 0.5,
        linetype = "dotted")
```

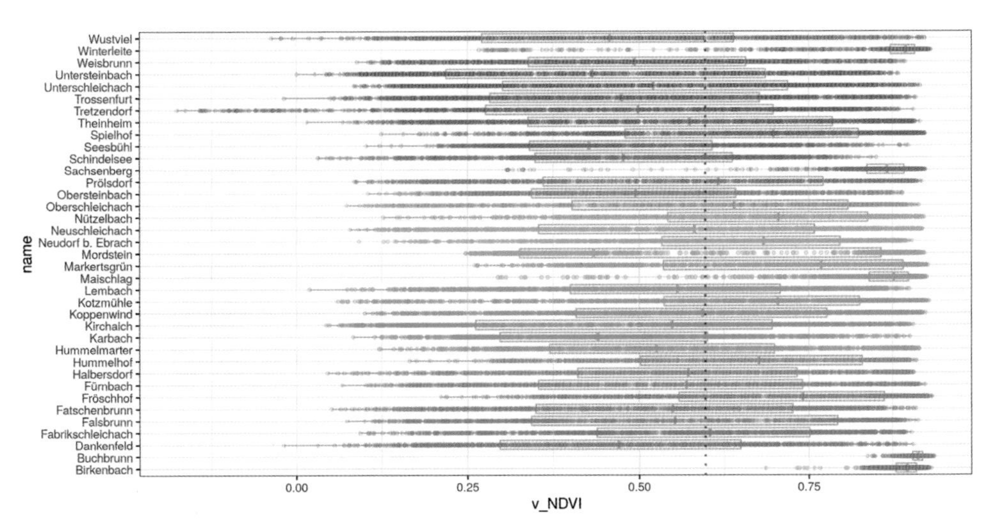

Figure 14.8 Box plot of all NDVI values including mean NDVI values across all samples. NDVI, normalized difference vegetation index.

TASK

Omit or add new *ggplot2* functions and check what has changed. Search for more plot options and try to implement them, for example, `geom_bar()`, `geom_density()` or `geom_violin()` instead of `geom_boxplot()`.

Further options can be added, for example, a different colour definition for the points, such as gradual values using the slope column or splitting the plot by different groups using the `facet_grid()` function (Figure 14.9):

```
ggplot(df, aes(name, v_NDVI, col = v_slp)) + geom_
boxplot(color = "grey", outlier.shape = NA) +
    geom_point(size = 1, alpha = 0.2) + facet_grid(cols =
vars(v_class)) + coord_flip()
```

The box plot shows that NDVI values across different locations differ between the land cover classes 'forest', 'meadow' and 'soil/rock' and that certain locations have a steeper slope (blue points). In some cases, only a subset of the locations should be displayed to highlight a certain pattern. Thus, we subset the data set for three locations based on their name:

```
df.s <- df[df$name %in% c("Birkenbach", "Fabrikschleichach",
"Untersteinbach"),
    ]
```

The new data frame contains the different environmental information for the three specified locations; these can now be plotted again as shown earlier (Figure 14.10). We then save the plot to a variable p and print p to display the plot:

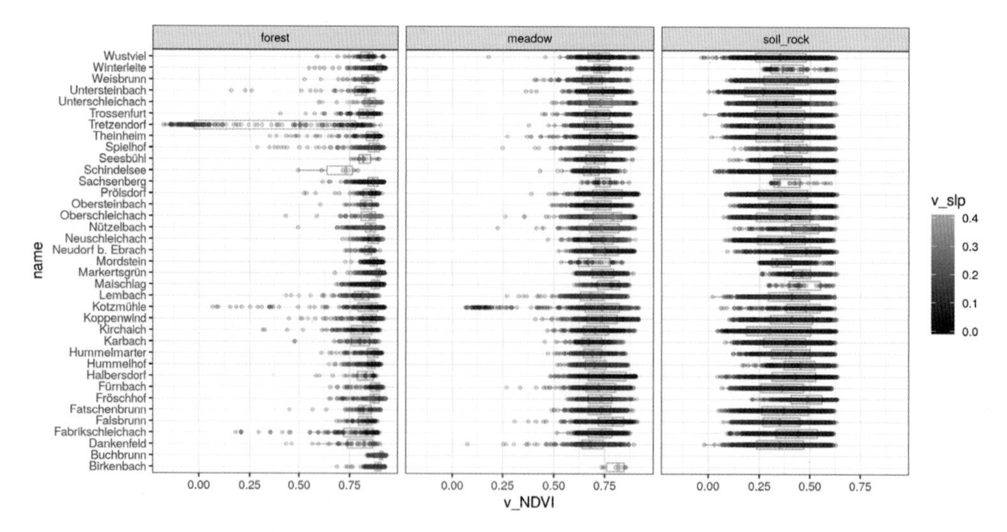

Figure 14.9 Box plot of NDVI values split by group of land cover classes.

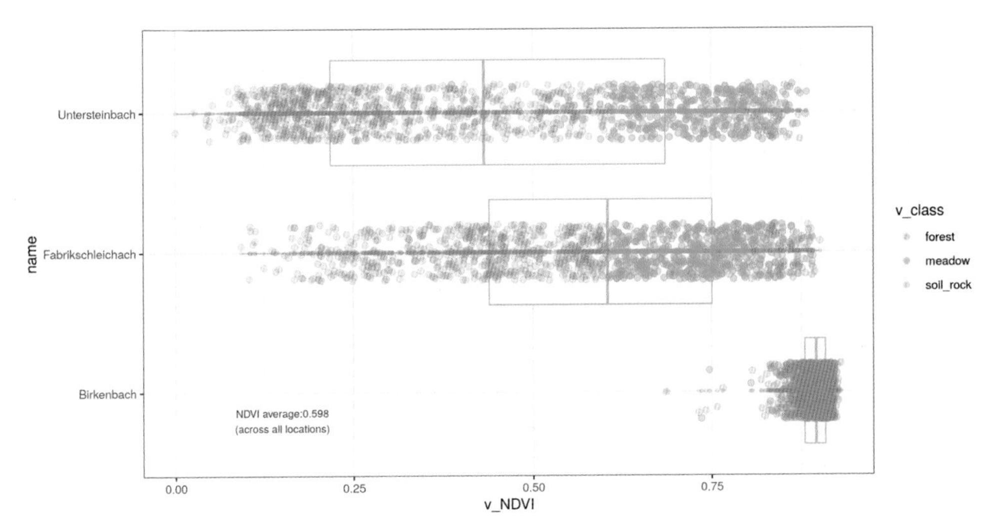

Figure 14.10 Subset of the NDVI values for selected locations shown as a box plot with several design adaptations.

```
p <- ggplot(df.s, aes(name, v_NDVI, col = v_class)) + geom_
boxplot(color = "grey",
    outlier.shape = NA) + coord_flip() + geom_point(size = 1,
alpha = 0.2) +
    geom_jitter(size = 2, alpha = 0.25, width = 0.2) +
annotate("text", x = 0.8,
    y = 0.15, size = 2.7, color = "grey20", label =
glue::glue("NDVI average:{round(mean(df$v_NDVI),3)}\n (across
all locations)"))

p  # Call the saved plot
```

> **TASK**
>
> Change the order of the *ggplot2* functions, for example, move the `geom_boxplot()` to the bottom and evaluate how it changes your plot.

Through the coloured points, you can clearly see that certain locations are mainly covered with forest while others are predominantly soil and rocks. This is also reflected by the box plot position and width. To better display overlapping points, variable positioning (jitter) was set. Adding such modifications can be beneficial for visualizing the data but must be done with care and clearly stated to avoid any confusion. For this plot, we could also omit the `coord_flip()` function and display the names along the x-axis.

For presentations, a simple but potentially valuable extension to this plot is the subsequent animation of values for each village. This can be achieved using the *gganimate* package and by adding a few animation functions to your *ggplot*:

```
library(gganimate)
p + transition_states(name, transition_length = 2, state_
length = 1) + enter_fade() +
    exit_shrink() + ease_aes("cubic-in-out", interval = 1e-07)
```

> **INFO**
>
> Further *ggplot2* functions and examples can be found at https://ggplot2.tidyverse.org/reference/ or http://r-graph-gallery.com and many more online and offline resources.

14.3 Finalizing and saving the plot

The last step in the graph creation process is finalizing the plot for printing or presentation purposes. Depending on the aim, either titles or captions need to be added and/or colours need to be changed. For our example, we change the axis and legend labels and add a title and subtitle using the + `labs()` function. Additionally, we add colours that reflect land cover in a more meaningful way (Figure 14.11):

```
ggplot(df.s, aes(name, v_NDVI, col = v_class)) + coord_flip()
+ geom_point(size = 1,
    alpha = 0.2) + geom_jitter(size = 2, alpha = 0.5, width =
0.25) + geom_boxplot(color = "black",
    outlier.shape = NA, alpha = 0) + scale_color_manual(values
= c("darkgreen",
    "green", "brown"), labels = c("Forest", "Meadow", "Soil/
Rock"), name = "Land Cover") +
```

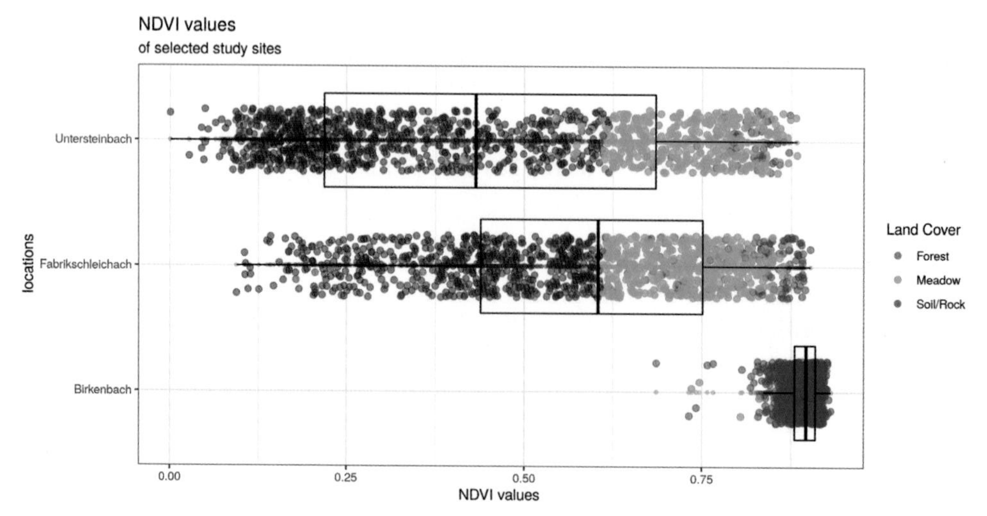

Figure 14.11 Adapted colour definitions and labels for the NDVI boxplot information.

```
    labs(title = "NDVI values", subtitle = "of selected study
sites", y = "NDVI values",
        x = "locations", col = "landcover")
```

Note that the `geom_boxplot()` function is moved below `geom_jitter()` and set to translucent so that the box plot outlines are visible on top of the points. Also remember that we applied `flip_coord()`; thus, the x-axis is located to the left and no longer on the bottom. Further options, such as text within the plot or arrows, could also be added. Saving the final plot to your hard drive is done using the `ggsave()` function. It saves the last plot that has been printed and allows us to define the file format or resolution:

```
ggsave("figure.pdf")
ggsave("figure_small.png", width = 6, height = 6, dpi =
"screen")
```

Usually, the dots per inch value for printing should be 300 dpi, which is the default. For presentations, a lower resolution is enough and reduces the file size (screen = 72 dpi). The width and height can also be adapted and do not need to be the same value. If an elongated graph is more appropriate, the width could be set to 10 while the height is kept at 6. Additionally, units and output format can be defined.

TASK

Export your plot in different formats and using different width/height settings. Check the resulting image to understand the implications.

14.4 Summary and further reading

Apart from graph making, data used for statistical analysis and statistical results can be reinserted into a spatial object. For example, we can run a linear regression of two variables and add the residuals back to the data frame and thus to the spatial object, which allows us to display residuals on a map. If you want to learn more about data visualization, we highly recommend the following books: *Data Visualization* by Kieran J. Healy (Princeton University Press) and *Fundamentals of Data Visualization: a Primer on Making Informative and Compelling Figures* by Claus O. Wilke (O'Reilly). Many more books, either about using R for plotting graphs or more general visualization options, exist. The graph making outlined in this chapter represents just a small fraction of the capabilities of *ggplot2*; much more sophisticated graphs can be created. A huge diversity of books and online resources, such as the R Graph Gallery (http://r-graph-gallery.com), provide more details on *ggplot2* graphs. The number of resources and options available within *ggplot2* are vast and can be overwhelming; however, mastering them will empower you to create highly informative and visually appealing graphs. The basic *ggplot2* options can be further expanded by using *ggplot2*-related packages, as well as packages aimed at animations, such as *ggrepel* for better labelling, *ggstatsplot* to add statistical information or *ggRadar* for radar plots.

15. Creating maps in R

Creating maps in R is not as straightforward as with QGIS. However, you can create graphs and maps in the same environment as your analysis. Once you have a working script for your favourite map design, reusing it and adapting it to new data is quite easy. In the long run, you have less work to do if you start creating maps with R. Maps such as the one shown in Figure 15.1 can look quite like QGIS maps but require a fair amount of coding.

Several packages in R have map plotting capabilities. The *tmap* and *cartography* packages can be used to create thematic maps. The *raster* and *stars* packages come with plot functions for raster data, while *sf* is the package of choice for vector elements in maps created with R. In the following sections, we use *ggplot2, sf, ggspatial, RStoolbox* and other packages that make it easy to create maps in R. This will be a simple task once all the necessary functions are known to you and you have understood them. We use the *ggplot2* family because the functions provided by *RStoolbox, sf* and *ggspatial* can all be combined with *ggplot2*.

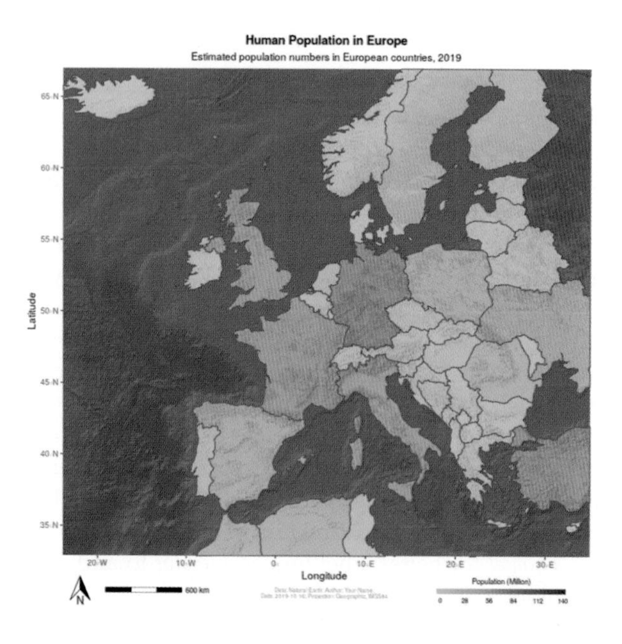

Figure 15.1 Example map of Europe using data like those employed in the QGIS map examples. Data and functions used are introduced below.

15.1 Vector data

We start with a simple map of Europe using Natural Earth Data (https://www.naturalearthdata.com). First, we must install and activate the respective packages:

```
library(rnaturalearth)
library(rnaturalearthdata)
library(sf)
```

Then, we can download medium-resolution country data and plot it:

```
world <- ne_countries(scale = "medium", returnclass = "sf")
plot(world)
```

The different plots (Figure 15.2) already show which data are available in this object. We can look at the attributes of this vector object in more detail by calling:

```
View(world)
```

Apart from the country names, continent association and income or development status are also available. We are interested in a map of Central Europe where our study area is located; thus, we subset our vector to the European continent:

```
europe <- world[world$continent == "Europe", ]
```

After plotting the initial map, we can see how the subsetting has worked out (Figure 15.3). The result shows all countries that are flagged as European in the attribute table:

```
plot(europe[1])
```

However, the area is still too large to outline the location of our study area; thus, we decide to subset by region and remove Russia:

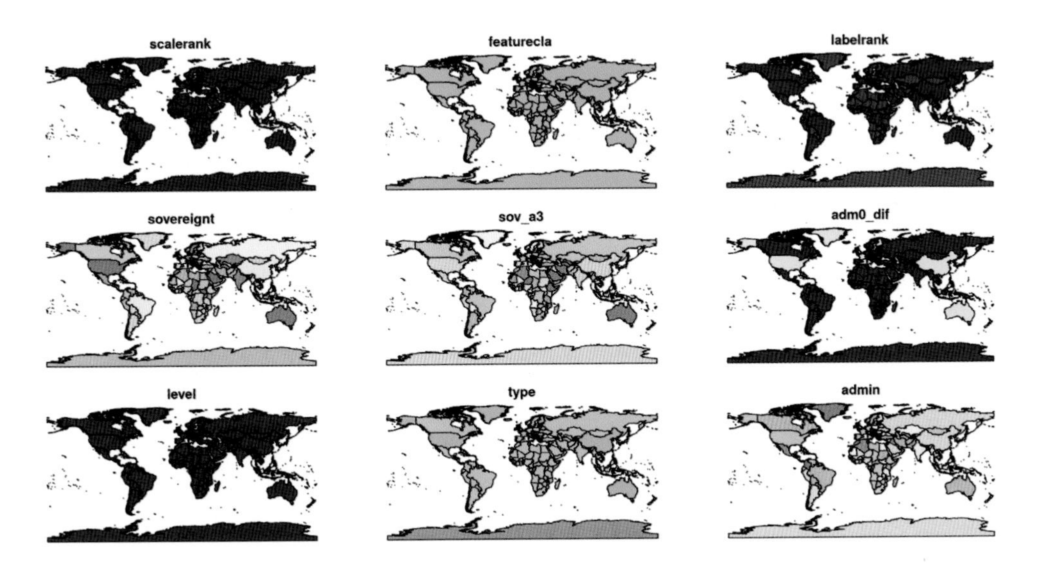

Figure 15.2 World maps displaying different information.

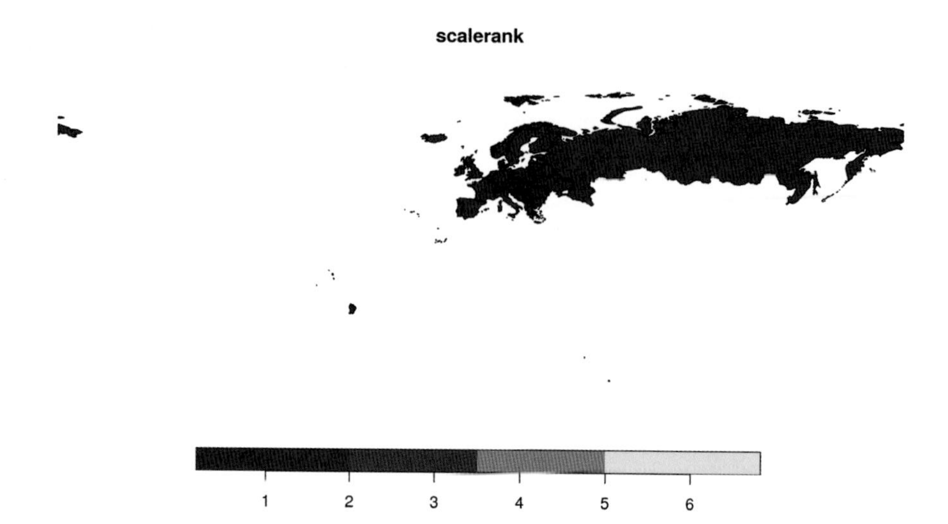

Figure 15.3 Map of all spatial features with the '`continent == "Europe"`' definition.

Figure 15.4 Improved subsetting of the world map to the European continent.

```
europe <- world[world$continent == "Europe" & world$name !=
"Russia", ]
```

The '`&`' sign combines two queries; == queries equal values while != means `unequal`. Plotting the result shows that we are slowly approaching our goal of a Central European map (Figure 15.4):

```
plot(europe[1])
```

Subsetting it even further might be possible by adding more statements, but other options are available when we apply the *ggplot2* functionality (Figure 15.5). Plotting the data without any further refinements can be done by:

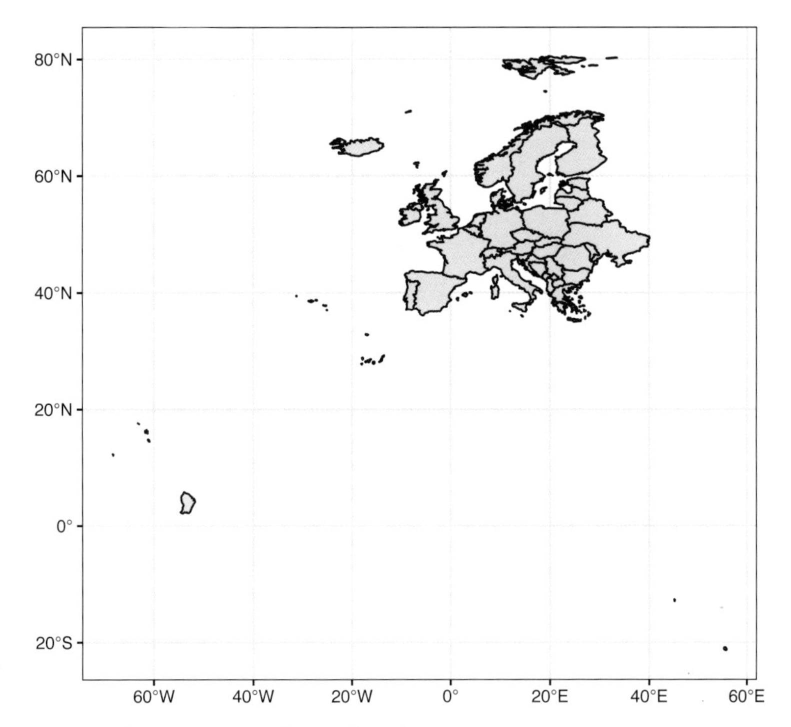

Figure 15.5 Map of Europe using the ggplot2 functions.

```
library(ggplot2)
ggplot(data = europe) + geom_sf()
```

Within *ggplot2*, we can add a `coord_sf()` function to limit the x and y extents of our plot, which allows us to zoom in to continental Europe (Figure 15.6):

```
ggplot(data = europe) + geom_sf() + coord_sf(xlim = c(-12,
40.12), ylim = c(34,
    72), expand = FALSE)
```

The values added into the `xlim` and `ylim` arguments can be copied from the map coordinates shown on the map. Adding a title and including functions to automatically extract information from the attribute table, such as the number of unique country names, can also be done using `ggtitle()` (Figure 15.7):

```
ggplot(data = europe) + xlab("Longitude") + ylab("Latitude") +
geom_sf() + coord_sf(xlim = c(-12,
    40.12), ylim = c(34, 72), expand = FALSE) + ggtitle("Map
of Europe", subtitle = paste0("(",
    length(unique(europe$name)), " countries)"))
```

Adding attributes from the table, such as colours, to the map can be achieved by adding within the `geom_sf()` function a `fill=` part and assigning the name of an attribute column (Figure 15.8). In this example, we use the population estimate (`pop_est`):

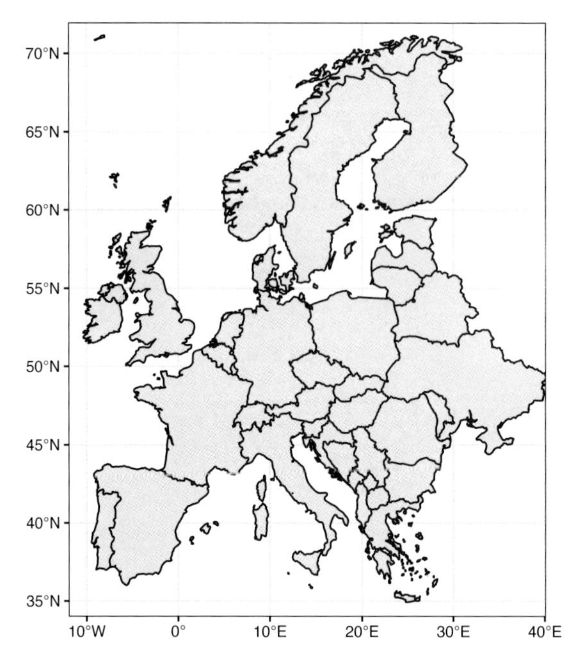

Figure 15.6 Map of Europe with reduced extent.

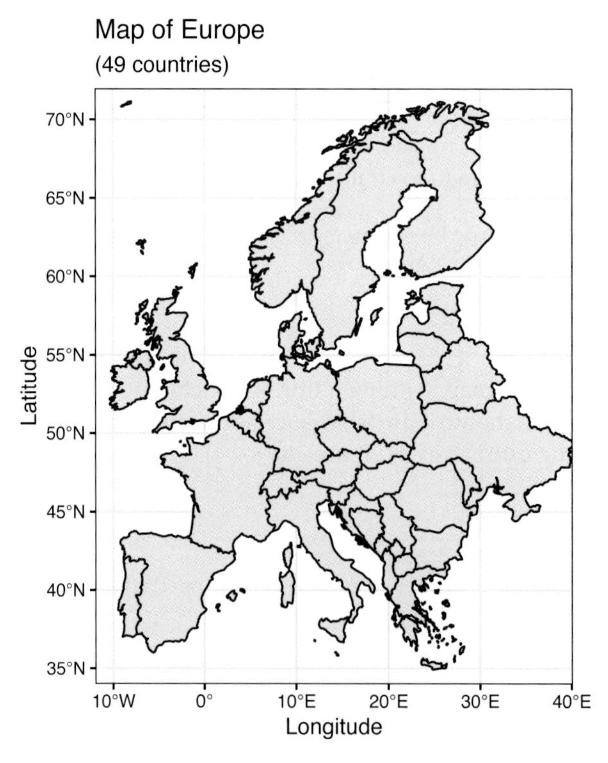

Figure 15.7 Further modified map of Europe, including a title, with automatic extraction of statistics.

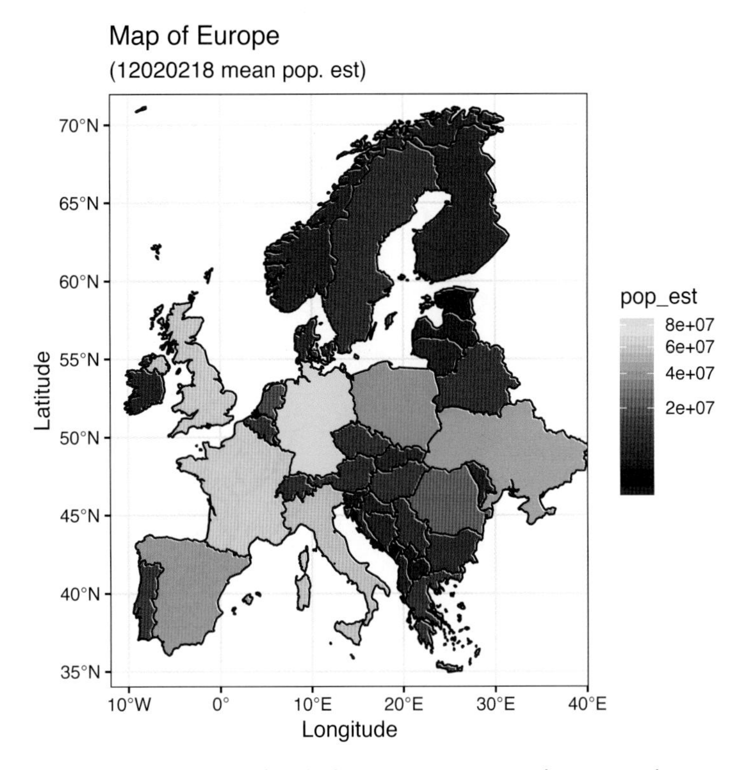

Figure 15.8 European countries mapped with their respective population numbers.

```
ggplot(data = europe) + xlab("Longitude") + ylab("Latitude") +
geom_sf(aes(fill = pop_est)) +
    coord_sf(xlim = c(-12, 40.12), ylim = c(34, 72), expand =
FALSE) + ggtitle("Map of Europe",
    subtitle = paste0("(", round(mean(europe$pop_est)), " mean
pop. est)")) +
    scale_fill_viridis_c(option = "plasma", trans = "sqrt")
```

> ## TASK
>
> Change the map projection by adding '`+ coord_sf(crs = "+init=epsg:3035")`' to your existing *ggplot*. Play with the attributes that are plotted to create maps displaying different measures. Alter the *ggplot2* code based on what you have already learnt to see how the map changes.

Colouring the plot is defined by the `scale_fill_viridis_c()` function where colour ramps can also be modified further or values transformed. What is still missing on the map, which you might remember from Chapter 4, are an arrow indicating north and a scale bar (Figure 15.9). These can be added using the *ggspatial* package and two new functions, namely `annotation_scale()` and `annotation_north_arrow()`:

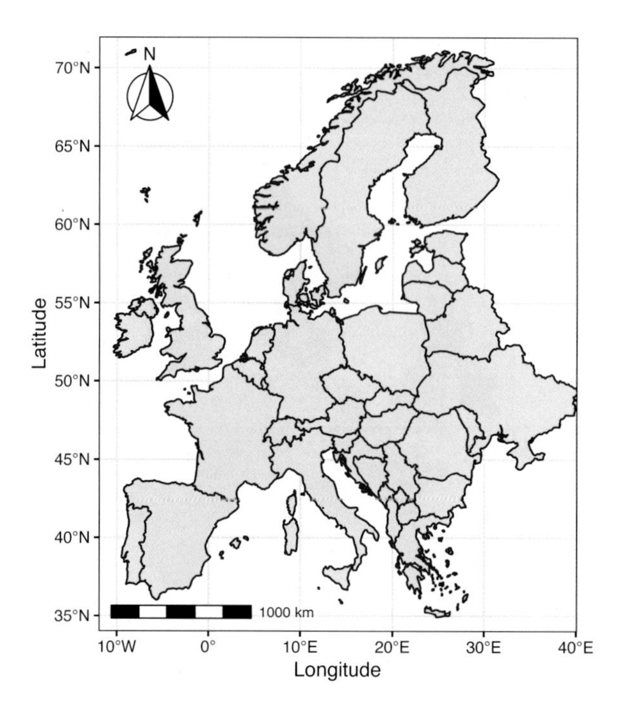

Figure 15.9 Relevant elements added to the map of Central Europe. The scale bar and arrow indicating north can be adjusted further.

```
library(ggspatial)
ggplot(data = europe) + xlab("Longitude") + ylab("Latitude") +
geom_sf() + annotation_scale(location = "bl",
    width_hint = 0.5) + annotation_north_arrow(location =
"tl", which_north = "true",
    style = north_arrow_fancy_orienteering) + coord_sf(xlim =
c(-12, 40.12),
    ylim = c(34, 72), expand = FALSE)
```

The location details `bl` or `tl` indicate *bottom left* and *top left*, respectively. Additionally, the names of the countries can be included using `geom_text()`, and several additional visual elements could be inserted. The design can be changed further using additional functions; consult the respective manuals of these packages to learn how to use these functions.

15.2 Plotting study area data

We can create similar maps for our specific study area and explore any mapping options further. For this example, we use the settlement vector object called 'places'. For this object, we may face the challenge of overlapping labels of settlement names because some settlements can be close to one another. Such issues must be checked and dealt with. First, import the settlement vector object unless you have already done so:

```
places <- st_read("place_UTM_WGS84_32N_clip.shp")
```

Now, we can plot this vector; using the package *ggrepel*, which automatically positions non-overlapping text, we can also add all villages names (Figure 15.10):

```
library(ggrepel)

ggplot(data = places) + xlab("Longitude") + ylab("Latitude") +
geom_sf(fill = "lightgray",
    color = "gray") + geom_text_repel(data =
data.frame(st_coordinates(places)),
    aes(X, Y, label = places$name), size = 3, fontface =
"bold") + annotation_scale(location = "bl",
    width_hint = 0.5) + annotation_north_arrow(location =
"tl", which_north = "true",
    style = north_arrow_fancy_orienteering)
```

The map shows that labels do not overlap each other. If it were not possible to position labels such that they did not overlap, the function would automatically move them to a free space and add a connecting line. This map is not currently displaying the results of the spatial analysis we conducted previously. To make the map more meaningful, we must insert background raster data, such as the normalized difference vegetation index

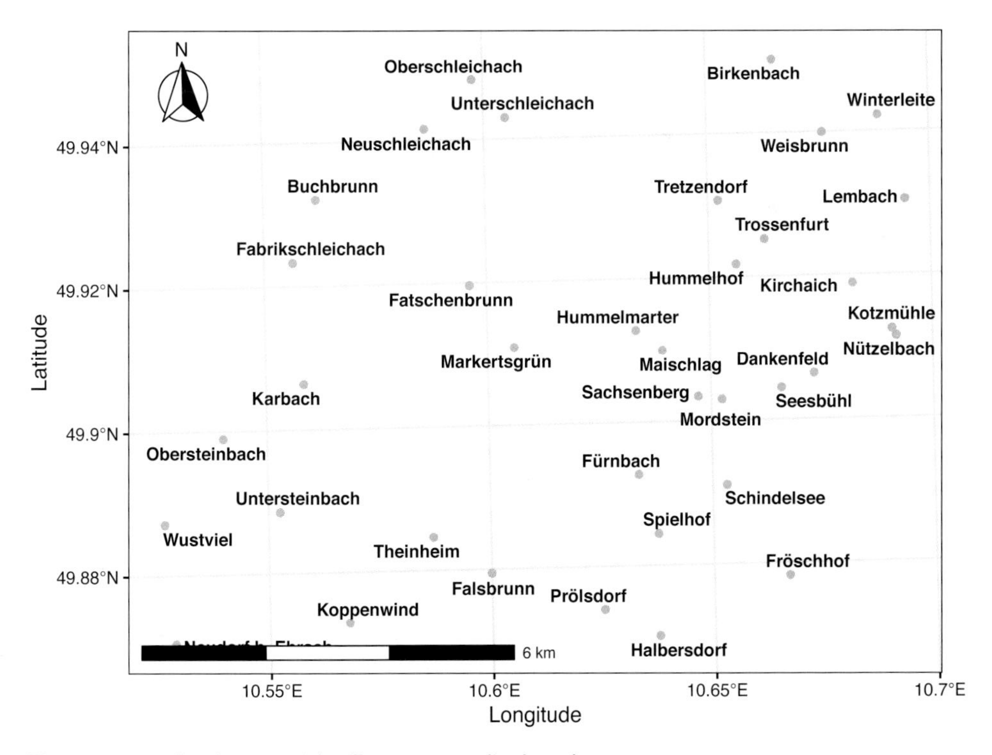

Figure 15.10 Study area with village names displayed.

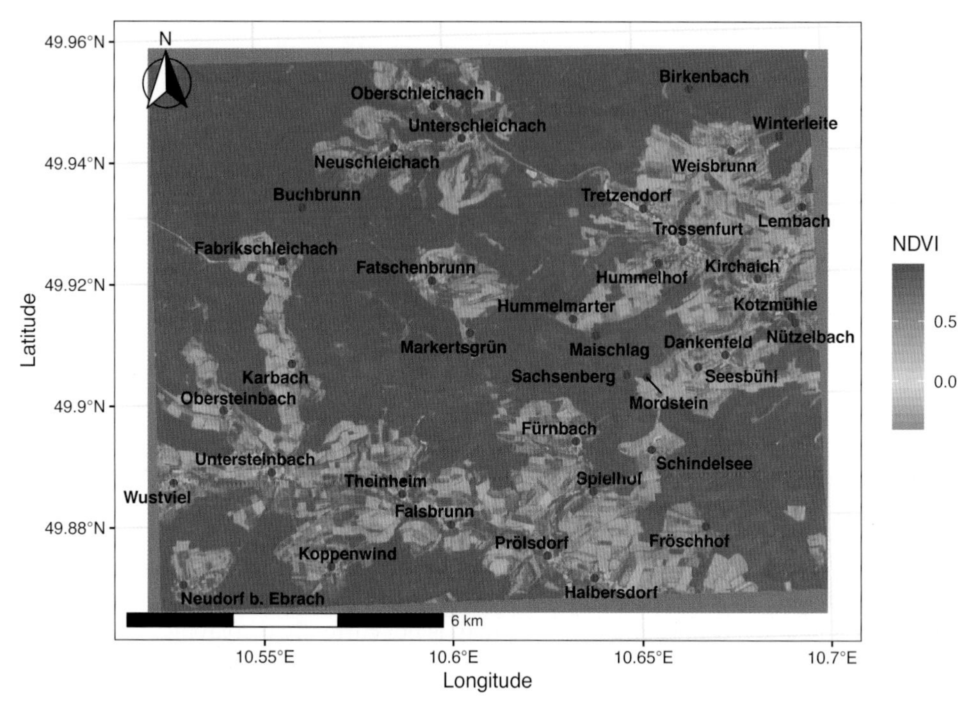

Figure 15.11 NDVI mapped for the study area, including elements such as names and scale bar. NDVI, normalized difference vegetation index.

(NDVI) or the classification (Figure 15.11). First, we import the classification and NDVI data set used in the previous chapters. Then, we plot them together with the vector data we used previously.

The *RStoolbox* package provides a raster plotting function, namely `ggR()` for single rasters and `ggRGB()` for multiband rasters. These can be combined with the mapping code we introduced earlier:

```
library(RStoolbox)

ggplot(data = places) + ggR(ndvi, geom_raster = TRUE, ggLayer
= TRUE, alpha = 1) +
    scale_fill_gradient2(low = "blue", mid = "lightgray", high
= "darkgreen",
        name = "NDVI", midpoint = 0) + xlab("Longitude") +
ylab("Latitude") +
    geom_sf(fill = "lightgray", color = "blue") + geom_text_
repel(data = data.frame(st_coordinates(places)),
    aes(X, Y, label = places$name), size = 3, fontface =
"bold") + annotation_scale(location = "bl",
    width_hint = 0.5) + annotation_north_arrow(location =
"tl", which_north = "true",
    style = north_arrow_fancy_orienteering)
```

The only difference to the previous code is the `ggR()` function and the corresponding colour gradient function, namely `scale_fill_gradient2()`. Data and colour are defined with these functions. Please note that the order is important. Moving the `ggR()` function to the bottom of the *ggplot2* will result in the raster being on top of the vector objects; thus, only the raster can be seen. An alternative is to move it to the bottom of the code and change the translucency (`alpha`) to 0.3, for example.

The same can be done for categorical raster data. In contrast to the `plot()` result, it creates a categorical legend (Figure 15.12):

```
ggplot(data = places) + ggR(sc_studyArea, geom_raster = TRUE,
ggLayer = TRUE,
    alpha = 1) + scale_fill_manual(values = c("darkgreen",
"green", "brown")) +
    xlab("Longitude") + ylab("Latitude") + geom_sf(fill =
"lightgray", color = "blue") +
    geom_text_repel(data = data.frame(st_coordinates(places)),
aes(X, Y, label = places$name),
        size = 3, fontface = "bold") + annotation_
scale(location = "bl", width_hint = 0.5) +
    annotation_north_arrow(location = "tl", which_north =
"true", style = north_arrow_fancy_orienteering)
```

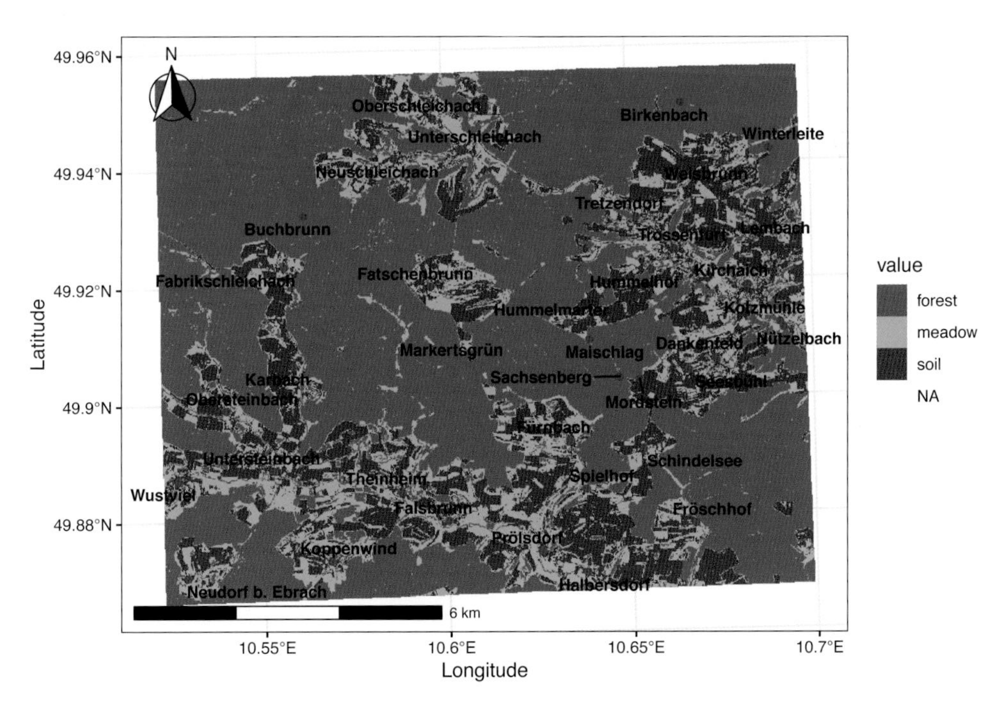

Figure 15.12 Land cover classification information included in our map using a categorical legend.

Within the `scale_fill_manual()` function, you can also define other colours for each class; the `geom_text_repel()` function allows to adapt the text further, for example, with boxes or different colours. Optionally, we can access further raster background data through the *Natural Earth* data package:

```
rst <- ne_download(scale = 50, type = "MSR_50M", category =
"raster", destdir = getwd())
```

The spatial resolution of these global maps can be used as the background for large-scale maps, but it will not be visualized smoothly at a local scale. Our final map can be saved in various formats and resolutions:

```
ggsave("map.pdf")
ggsave("map_web.png", width = 10, height = 10, dpi = 300)
```

15.3 Summary and further reading

In this chapter, you learnt how to create maps in R. The *ggplot2* mapping options are only the start. These functions can be used to create the first map in this chapter (see Figure 15.1). However, more sophisticated options are available. Overview maps inside another map, as well as zooming in to maps outside the main map or arrows linking to plots can all be achieved in R. Checking Web pages such as *Making Maps with R* by Robin Lovelace, Jakub Nowosad and Jannes Muenchow (https://bookdown.org/robinlovelace/geocompr/adv-map.html) or *Drawing Beautiful Maps Programmatically with R, sf and ggplot2 – Part 1: Basics* by Mel Moreno and Mathieu Basille (https://www.r-spatial.org/r/2018/10/25/ggplot2-sf.html) will provide you with much more information about map making in R.

TASK

Explore the capabilities of the R packages *rayshader* (https://github.com/tylermorganwall/rayshader) and *geoviz* (https://github.com/neilcharles/geoviz) to display spatial data three-dimensionally.

Afterword and acknowledgements

We hope you have enjoyed reading this book and have learnt how to get started with spatial data analysis. This book ends where *Remote Sensing and GIS for Ecologists* (http://book.ecosens.org) began, that is, with (more advanced) spatial analysis in R and a deeper insight into remote sensing methods for ecology. If you want to expand your knowledge further, we recommend continuing with that book.

We thank several people who helped us during the writing process. Several colleagues, including Alexandra Bell, Asja Bernd, Yrneh Ulloa, Malin Fischer, Thorsten Dahms and Patrick Sogno, helped us by reviewing the manuscript, providing feedback on the data used and critically appraising the explanations of the different approaches. Finally, many students provided feedback during our courses over the years, which formed the basis for this book.

This book would not have been possible without the continuous development of open-source software by the open-source community, especially the QGIS and R software developers. We are very grateful to this community for providing professional, freely available software packages that are crucial to reproduce scientific approaches. It empowers scientists around the world to pursue different research projects and strengthens collaborations.

We thank Nigel Massen for the opportunity and support to publish this book and for his valuable feedback and encouragement.

Feedback

Any feedback is greatly appreciated. Please go to our Web page http://book.ecosens.org and fill out the feedback/Get in Touch form for any feedback regarding typos, mistakes or general ideas on how to improve the book. We are happy to incorporate as many suggestions as possible in future editions.

Open-source software

Most tasks shown in this book can be easily achieved with QGIS and R. However, some additional functions might not (yet) be implemented or their performance is not good enough. Please do not hesitate to submit any requests, bug reports or even new functions to the respective software environments. Due to the fast developmental cycle, many new features become available within a short time frame, so keep an eye out for these new developments. Other software packages that are useful for ecologists working in a spatial context exist and might provide different or more sophisticated methods for your analysis. These include:

- System for Automated Geoscientific Analyses (SAGA) GIS (http://www.saga-gis.org/en/index.html): this is a comprehensive stand-alone GIS with a modular plug-in structure. SAGA provides a comprehensive suite of functions for terrain analysis, hydrological modelling and geostatistics. SAGA functions can be accessed from QGIS.
- gvSIG Association (http://www.gvsig.org/en/web/guest): a comprehensive stand-alone GIS developed for administrative purposes. It is well suited to interactive data exploration and analysis.
- User-friendly Desktop Internet GIS (uDig) (http://udig.refractions.net): stand-alone user-friendly GIS that focuses on online data sources.
- Orfeo ToolBox (https://www.orfeo-toolbox.org): Orfeo ToolBox is remote sensing image processing software (stand-alone or Application Programming Interface) from the French National Centre for Space Studies (Centre national d'études spatiales). Modules are available from QGIS.
- Opticks (https://opticks.org): stand-alone remote sensing image processing software for multispectral and hyperspectral imagery.
- PolSARPro (https://earth.esa.int/web/polsarpro/home): radar processing software from the European Space Agency.
- GRASS GIS (https://grass.osgeo.org): GRASS or GRASS GIS is a hybrid software combining GIS and remote sensing functionalities. Its acronym stands for Geographic Resources Analysis Support System and it is one of the largest and most powerful raster manipulation open-source systems available. GRASS is based on a sophisticated database, which is well adapted to multi-user environments.

Many more are available. To find out more, go to the Open Source Geospatial Foundation (https://www.osgeo.org).

References

Estallo, E.L., Ludueña-Almeida, F.F., Visintin, A.M., Scavuzzo, C.M., Lamfri, M.A., Introini, M.V., Zaidenberg, M. and Almirón, W.R. (2012) Effectiveness of normalized difference water index in modelling *Aedes aegypti* house index. *International Journal of Remote Sensing* 33 (13): 4254–4265.

Huete, A., Didan, K., Miura, T., Rodriguez, E.P., Gao, X. and Ferreira, L.G. (2002) Overview of the radiometric and biophysical performance of the MODIS vegetation indices. *Remote Sensing of Environment* 83 (1–2): 195–213.

Huete, A.R. (1988) A soil-adjusted vegetation index (SAVI). *Remote Sensing of Environment* 25 (3): 295–309.

Jordan, C.F. (1969) Derivation of leaf-area index from quality of light on the forest floor. *Ecology* 50 (4): 663–666.

Lillesand, T.M. and Kiefer, R.W. (1987) *Remote Sensing and Image Interpretation*. New York, NY: John Wiley and Sons.

Qi, J., Chehbouni, A., Huete, A.R., Kerr, Y.H. and Sorooshian, S. (1994) A modified soil adjusted vegetation index. *Remote Sensing of Environment* 48 (2): 119–126.

Rouse, J.W., Haas, R.H., Schell, J.A. and Deering, D.W. (1974) Monitoring vegetation systems in the Great Plains with ERTS. In S.C. Freden, E.P. Mercanti and M. Becker (eds) *Third Earth Resources Technology Satellite–1 Symposium*. Volume I: Technical Presentations. NASA SP-351. Washington, DC: NASA. pp. 309–317.

Tucker, C.J. (1979) Red and photographic infrared linear combination for monitoring vegetation. *Remote Sensing of Environment* 8 (2): 127–150.

Xu, H. (2006) Modification of normalised difference water index (NDWI) to enhance open water features in remotely sensed imagery. *International Journal of Remote Sensing* 27 (14): 3025–3033.

Index

Page numbers in **bold** indicate tables and in *italic* indicate figures.